Contents

Pigeon Poo, the Universe & Car Paint

and other awesome science moments

Dr

KARL KRUSZELNICKI'S

new moments in science # 1

Illustrations by Reg Lynch and Peter Pound

HarperCollinsPublishers

HarperCollins*Publishers*

First published in Australia in 1996
Reprinted in 1996, 1997
by HarperCollins*Publishers* Pty Limited
ACN 009 913 517
A member of the HarperCollins*Publishers* (Australia) Pty Limited Group

HarperCollins*Publishers*
25 Ryde Road, Pymble, Sydney, NSW 2073, Australia
31 View Road, Glenfield, Auckland 10, New Zealand
77-85 Fulham Palace Road, London W6 8JB, United Kingdom
Hazelton Lanes, 55 Avenue Road, Suite 2900, Toronto, Ontario M5R 3L2
and 1995 Markham Road, Scarborough, Ontario M1B 5M8, Canada
10 East 53rd Street, New York NY 10032, USA

National Library of Australia Cataloguing-in-Publication data:

Kruszelnicki, Karl, 1948– .
Pigeon Poo, the universe and car paint.
ISBN 0 7322 5723 9.
1. Science–Popular works. I. Title. (Series: New moments in science; 1).
500

Cover illustration by Reg Lynch
Cartoons by Reg Lynch
Technical illustrations by Peter Pound
Printed in Australia by Griffin Paperbacks, Adelaide

9 8 7 6 5 4 3 97 98 99

THANKS

Whenever the medieval cathedral-builders erected a cathedral, they deliberately incorporated a 'mistake' into it, to show that only God is perfect. For example, one side might have two pointed towers, while the other side might have one pointed tower and one blunt tower. I assume that there will be some mistakes in this book, and I hereby take the blame for all of them. The first people to point out each mistake will get a free copy of my next book.

I would like to dedicate this book to the many people that made it possible. Fitzroy Boulting (who should have his own 0055 number) started this project with HarperCollins. Angelo Loukakis was the godfather, while Jude McGee (famous singer from Pel Mel) was the hands-on mother (and I'm sorry the manuscript was a month late, but I had a really good time in the USA). Louise McGeachie is a wonderfully artistic designer, and I was lucky to get her. Caroline Pegram has combined the various skills of research assistant, paintbox operator, audio-visual desk mixer and personal assistant into an indescribable job description, and has worked at the 'coal face' on this book, pulling together information, turning electro-magnetic waveforms into hard copy, and solidifying concepts.

Dan Driscoll (Producer, ABC National Programs Unit, NPU) has been a sensitive and insightful nurturer of these concepts, and has helped advance them further. Paul Vadasz, as head of the ABC NPU, provided the environment in which these stories first appeared as short radio stories called 'Great Moments in Science', once per week, all over Australia. Without the radio stories, there would not have been any book stories.

I would especially like to thank Joel Harp, Michael Buckland, Glen Croxson, Penny Clay, Steve Blaber, Martha Morrow, Ted Ankrum and Nature magazine. Reg Lynch and Peter Pound (Caroline's brother-in-law, but if you can't do nepotism with your family and friends, who can you do it with?) provided the wonderful illustrations.

Firstly, and finally, Mary Dobbie and our children helped polish these stories at many stages in their development.

KARL KRUSZELNICKI

PIGEON POO, THE UNIVERSE & CAR PAINT

Some people call pigeons 'flying rats'. It's not a very nice name. These people probably don't know that pigeons have a lot going for them. They are the fastest racing animal, and they feed their babies a milk that is very similar to human milk. Pigeons landed a couple of scientists a Nobel Prize, because they bothered to find the difference between Pigeon Poo and The Beginning of The Universe. Even better, pigeons have got their revenge on the car manufacturers for contributing to acid rain.

History of pigeons

The earliest pigeon that we know of is a fossilised pigeon, *Gerandia calcaria*, that lived in what is now France, between 23 million and 17 million years ago.

There are about 250 to 300 species of birds in the pigeon family. They live worldwide, but prefer the warm climates. They usually have deep-chested bodies with short legs, short necks and small heads. Their long wings and powerful chest flight muscles make them very swift, and strong, fliers. They have an unusual saddle of skin (called the 'cere') between their bill and their forehead. They also have a very characteristic bobbing of the head when they walk about. Another special feature is that pigeons can 'suck' liquids, thanks to their muscular necks. No other bird can do this — they first have to sip a little water, then raise their beaks out of the water, and then finally, with a little help from gravity, swallow.

Pigeons like to eat fruit, nuts, and seeds. They call with a very characteristic 'cooing' sound. They usually pair up for life, and if one of them dies, the surviving pigeon will be very slow to take a new mate. The female pigeon usually lays one or two eggs each year.

In this family of 250–300 different species, we call the big ones 'pigeons', and the little ones 'doves', but size is the only real difference between them. However, the 'Dove of Peace' is really the white domesticated pigeon.

Pigeons were probably the first bird to be domesticated by humans. Pigeons can be seen in Mesopotamian figures, coins and mosaics dating back some 6,500 years. They were tamed at least 5,000 years ago by the Egyptians, who ate them. Written records from the 5th Egyptian Dynasty, dating back to 3,000 BC, discuss the domestication of the pigeon.

Message carriers with wings

But 2,000 years ago, the ancient Romans and Greeks found another use for pigeons — carrying messages. In fact, pigeons carried news about the original Olympic Games in Ancient Greece. Around 1150 AD, the Sultan of Baghdad set up a regular Pigeon Post. In 1849, the telegraph service between Berlin and Brussels was interrupted for a short while — but pigeons kept the messages flowing. And even though the last two world wars were hi-tech, the armed forces still used homing pigeons.

Homing pigeons have fantastic navigation skills. One pigeon released by the US Army Signal Corps made its way back home over a distance of some 3,700 kilometres. They navigate by listening to infrasound (the low-frequency sound of waves beating on the beach, or wind rumbling past mountains), by taking

THE HORMONES OF PIGEONS

Hormones have odd effects in pigeons. The mere act of a female seeing her mate build a nest will lead to a whole series of changes. Her neuro-endocrine system is stimulated so that the level of the female sex hormone, oestrogen, rises in her blood. This raised level of oestrogen makes her oviduct develop about five days before her eggs will eventually be laid!

Just the mere act of seeing another bird sitting on eggs, or she herself sitting on her own eggs, will cause an increase in the level of the hormone prolactin in her blood.

The increased level of prolactin starts the process of production of 'pigeon milk' from her crop. This production is timed so that the maximum amount of milk will be available once the eggs hatch. This means that the parent pigeons don't have to leave the nest when the baby pigeons are most helpless.

Once the eggs have hatched, the production of pigeon milk gradually drops, and this forces a situation where the young have to eat more adult-type food.

bearings on the Sun, by looking at ultra-violet light in the sky, by looking at polarised light from the Sun in the sky, by looking at landmarks on the ground, and by following the Earth's magnetic field (thanks to tiny magnets buried in their tiny brains).

When the Earth's magnetic field gets a temporary shake-up, pigeons can get confused. There's the famous case in northern France, back on 7 July 1988, when 3,000 homing pigeons were released just two days after a solar flare (an enormous explosion on the Sun). The sub-atomic particles thrown off by this explosion reached the Earth just as the pigeons were being released. The solar flare severely disrupted the Earth's magnetic field. Most of the pigeons were never seen again.

Under normal conditions, homing pigeons make it back to home base, flying at around 72 kilometres per hour. Racing pigeons, the Ferraris of Birdland, have been clocked travelling at 177 kilometres per hour — but that was with a bit of tail wind!

In America, there were once vast numbers of passenger pigeons. In the early 1800s, it was common to see migrating flocks made up of millions of birds. The flocks would be 5 kilometres wide, and 500 kilometres long, and would darken the sky for hours or days at a time. But passenger pigeons are now extinct — the last one died in the Cincinnati Zoological Garden, on 1 September 1914. They made us appreciate how terribly easy it is to drive a species to extinction! In fact, there is a monument to the passenger pigeon, in Wyalusing State Park, in Wisconsin. The inscription on the monument says: 'This species became extinct through the avarice and thoughtlessness of man.'

WHAT WAS THE BIG BANG?

About 15 billion years ago, there was the time before the Big Bang. Our current science can tell us nothing about this time.

At one point in time, all the 'stuff' in the 'Universe' was concentrated into a small lump, smaller than a basketball.

Suddenly, the Big Bang happened, and this small lump began to expand. For a little while, the boundaries of our early Universe actually expanded faster than the speed of light (according to the Super Inflationary Theory). Einstein says that nothing can travel faster than the speed of light, so how can this be? Well, nothing can travel faster than the speed of light, in our Universe. Whatever the boundaries of our early Universe were expanding into, it was not our Universe.

The Universe was very hot, with temperatures of the order of millions of millions of millions of millions of millions of millions of degrees. It was much too hot for matter, as we know it today, to exist. Only pure energy could exist. With time, the Universe cooled down. Then sub-atomic particles 'condensed' out of the cooling energy, followed by atoms. Once atoms were around, matter as we know it could exist.

WHY THE BIG BANG THEORY IS CORRECT

The Big Bang Theory is believed by almost all the cosmologists (scientists who think about the origin of our Universe). There are three main pieces of evidence for the Big Bang Theory.

First, the most common elements in the Universe are hydrogen (about three-quarters) and helium (about one-quarter). The remaining elements make up a few per cent, at the most. This preponderance of hydrogen and helium was predicted by Ralph Alpher and Robert Herman in their 1950 paper, in *Review of Modern Physics*, entitled 'Theory of the Origin and Relative Abundance Distribution of the Elements'.

Second, there is a cosmic background radiation, at roughly the temperature predicted by Alpher and Herman in their 1949 paper in *Physical Review*, 'Remarks on the Evolution of the Expanding Universe'.

Third, the Universe is expanding, as a result of the remaining outward force provided by that first Big Bang.

Pigeon milk

In general, the only animals that feed milk to their babies are mammals — but pigeons (which are birds, not mammals) can make milk! Now, this pigeon milk does not come from a breast. Instead, it is made up from the cells that line the crop. The crop is part of the gullet (or oesophagus in humans). In humans, only females can normally make milk. But in the land of the pigeons, both the males and the females can deliver this milk. The stimulus for milk delivery is usually the act of the baby pigeon poking its beak down the parent's neck! The pigeon just sheds the cells from the crop, and mixes them with saliva, and feeds this cheesy–milky liquid to the babies. This 'pigeon milk' is very similar to the milk of mammals, with 15 per cent protein and 10 per cent fat.

By an amazing coincidence (perhaps it's *not* a coincidence), the same hormone that controls milk production in mammals also controls milk production in pigeons. This hormone is prolactin, and it causes the cells of the crop to grow and proliferate. In the early days of research into how prolactin affected milk production, laboratory pigeons were used.

Pigeon poo and cosmic background radiation

Even more amazing than pigeon milk is pigeon poo, a substance which led to a Nobel Prize!

In 1965, Arno A. Penzias and Robert W. Wilson of Bell Telephone Laboratories were trying to make phones quieter. As part of their research, they were measuring the

WHY THE BIG BANG THEORY IS INCORRECT - 1

The main reason the Big Bang Theory might be incorrect is that all the 'main pieces of evidence' for the Big Bang Theory have alternative explanations!

The percentages of elements in the Universe? Well before the Big Bang was even thought of, R. C. Tolman came up with similar figures. He published them in the *Journal of American Chemists*, in 1927, in a paper called 'Thermodynamic Treatment of the Possible Formation of Helium from Hydrogen'.

The predictions of the background temperature of the Universe? Way back in 1926, Arthur Eddington had predicted the background temperature, caused by the radiation of starlight, would be around 3 Kelvin.

And the expansion of the Universe? Some scientists claim that there is no expansion, but just a misinterpretation of the 'red shift' of the distant galaxies. Other scientists claim that the incredible timing and navigation apparatus associated with the Galileo spacecraft at Jupiter would allow measurement of the 'expansion of the Universe' on a local scale, inside our Solar System. They also claim that this 'expansion' has been looked for, but not found!

radio emissions coming from an enormous ring of gas that surrounds our galaxy, the Milky Way. They were also working with the first Telstar communication satellite. They wanted to reduce the background noise in the long-distance circuits, so they were testing these long-distance circuits by sending and receiving phone calls with a radio telescope. No matter how hard they tried, they couldn't get rid of all the background noise. Their measurements showed that the extra background radio noise seemed to come equally from all parts of the sky.

It so happened that the surface of their radio telescope was covered with pigeon droppings — which they called 'a white dielectric material'! Their co-workers said the background noise was just caused by the pigeon droppings, and that they should stop worrying about it. But Penzias and Wilson were good scientist–engineers. They 'walked that extra mile', and went to the trouble of removing the pigeon droppings — but the noise still remained!

So they asked Bernard Burke of the Massachusetts Institute of Technology, Boston, about their problem. Burke knew that a team of physicists, including Robert H. Dicke and P. J. E. Peebles, had been planning to look for a certain cosmic background radiation. He realised that the noise discovered by Penzias and Wilson could be this very cosmic background radiation. Burke put the two teams in contact with each other. The noise-that-wouldn't-go-away turned out to be the cosmic background radiation. In 1965, the two groups published papers that discussed the prediction, and the

> ## WHY THE BIG BANG THEORY IS INCORRECT - 2
>
> **There's another reason why the Big Bang Theory might be incorrect — the 'light-horizon' problem.**
>
> **The cosmic background radiation is an essential part of the currently accepted Big Bang Theory. This radiation is incredibly even and smooth, with any 'bumps' being less than one part per 100,000.**
>
> **Consider the cosmic background radiation coming to you, here on Planet Earth. Some of this radiation comes from your north, and some from your south — in other words, from opposite sides of the sky. These two separate radiations have reached you only in the last few moments. These radiations have not had the opportunity to 'talk' with each other, before they reached you — they were beyond each other's 'light horizons'.**
>
> **So how could these two radiations, from opposite parts of the sky, 'know' how to have the same temperature, accurate to one part in 100,000? There is currently no good answer for this question.**

discovery, of the cosmic background radiation.

So what exactly is this mysterious 'cosmic background radiation'? According to our current knowledge of cosmology, this noise is the Ancient Reverberation left over from the Titanic Explosion of the Big Bang when The Universe started up.

You've probably noticed that when you drop a rock into a pond, the big waves will eventually fade into tiny ripples. In the same way, the enormous temperatures of the giant explosion that started off The Universe have since cooled down to a cosmic background radiation that permeates The Entire Universe.

Various scientists had predicted that this radiation would exist, and that it would have a temperature of around 5 Kelvin (5 degrees above absolute zero). The actual temperature turned out to be around 2.735 Kelvin. The temperature of this radiation has been measured and found to be incredibly even across the sky — it's the same, no matter where you look, at least down to an accuracy of one part in 100,000.

Finding out that a certain background radio noise was not caused by Pigeon Poo on Arno's and Rob's telescope, but by the Beginning of The Universe, eventually led to Penzias and Wilson sharing the 1978 Nobel Prize for Physics.

Acid poo

Another odd thing about pigeon droppings is that they currently cause more damage to your car's paintwork than they did in the past. It's not because the paint is thinner, or of inferior quality. According to a report by the Ford Motor Company, the paint damage is worse because the bird droppings are now more acid than they used to be.

The worldwide acid rain crisis means that surface water (such as in lakes) is more acid than ever before. The birds drink this acidic water, and then excrete more acid in their droppings. So you should wash off the bird droppings as soon as possible. The longer they're left on the paint, the worse the damage will be. The car makers have had to reformulate their paints, to resist these very acidic droppings.

In heavily industrialised areas, such as Pennsylvania, the pigeon droppings are now a threat to steel bridges. When the acidic pigeon droppings are mixed with

ANOTHER BIG REVOLUTION IN PHYSICS?

Towards the end of the 19th century, it seemed as though all the Big Problems in physics had been solved. All that was left for scientists to do, it seemed, was to measure the properties of matter to a few more decimal places. There were a few problems that scientists 'swept under the carpet', but basically, the World of Physics seemed settled.

Suddenly, one of these problems erupted. The geologists had worked out a minimum age for the Planet Earth. But the engineers and physicists could not explain how the Sun could burn for such a long time if it were made of coal.

The answer was radioactivity, and it led to a revolution in physics. It led to nuclear weapons and nuclear medicine, X-rays and quantum physics, electronics and video recorders, space travel and global positioning system (GPS) units that give your position on the planet accurate to a few metres.

Robert H. Frisbee, the Technical Group Leader of the Advanced Propulsion Technology Group at the Jet Propulsion Laboratory in California, believes that a few 'inconvenient' problems exist today. These problems include the problems with the Big Bang (discussed in the boxes 'Why the Big Bang Theory is Incorrect'), the fact that over 90 per cent of the known Universe is 'missing', and that certain complementary sub-atomic particles can, under certain conditions, 'talk' to each other at much faster than the speed of light. He thinks that there is a reasonable chance that we are heading for another revolution in physics.

He would be very happy with such a revolution. It would probably make one of his projects, travelling at faster than the speed of light, easier to solve!

PIGEON TO THE RESCUE

When William Brew from Roswell, New Mexico, had his car break down in the desert, on a desolate road, kilometres from anywhere, he was not worried. He had a basket of pigeons that he was training to be homing pigeons. He used them to send a message to his wife, who called the Road Patrol Service.

the combination of industrial pollutants, salt from the roads, and acid rain, they form acids strong enough to eat through bridge girders! In the year 1985, Pennsylvania spent a total of $200,000 to clean up some 100 bridges near Pittsburg.

There is sulphur in petrol, and in coal. This sulphur comes out of the exhaust pipe and chimney stacks as sulphur dioxide gas, and eventually ends up as sulphuric acid in the clouds, causing acid rain — which affects us all. But in the animal kingdom, it's the so-called 'flying rats' that have alone managed to take their revenge!

REFERENCES

Did You Know?, Reader's Digest (Australia), Sydney, 1991, pp. 42, 103.
Encyclopaedia Britannica, 1996.
Exploring the Secrets of Nature, Reader's Digest Association Far East, 1994, pp. 66, 96, 202, 298, 379.
Keay, Colin, 'Big Bang (second round)', *the skeptic*, vol. 16, no. 1, Autumn 1996, pp. 51–52.
Nagy, Charles, 'Big Bang', *the skeptic*, vol. 15, no. 4, Summer 1995, p. 64.
'Pigeon power', *Sydney Morning Herald*, 18 August 1995, p. 28.

INSECT BALL OF FIRE

It might be a dog-eat-dog world out there in the business world, but it's even worse in the land of the insects. In Japan, the Giant Japanese Hornet is about four or five times bigger than the Japanese Honeybee — and it can kill 40 honeybees per minute. So, to stop being wiped out by the hornets, the honeybees have worked out a system of Neighbourhood Watch, which works on provocative dances and smells, backed up by the honeybees' very own Great Ball of Fire!

Solitary and social wasps

Now the hornet is basically just a large stinging wasp. There are two types of wasps. There are the 'solitary wasps', which mainly live by themselves, and are not very aggressive. Then there are the 'social wasps', which live in groups, and are easily provoked. Hornets have well-organised communities, with clearly defined roles for the females, the males, and the sterile workers.

Hornets build their nests out of a paper-like material. They make this material by chewing wood and the foliage of plants, adding lots of saliva, and then spitting out the gooey mixture. They usually wrap the nest in a paper envelope shaped like a football. In fact the word 'wasp' comes from the root wefan, which means 'to weave' and is a reference to the woven appearance of the paper nests. Often these nests have many storeys, or levels, which are hung underneath each other by paper ropes.

Hornets like a range of foods, including ripe fruit, insects and insect larvae. They are actually very much like bees, except a few times larger. They live in nests that range in size from a few hundred to 15,000 insects. Just like bees, adult hornets prefer carbohydrate foods, while young hornets like protein foods. Like bees, hornets are often attacked by bacteria and fungi. And hornets have similar work habits, and a similar caste system, to bees.

HORNETS

The Bald-faced (or White-faced) Hornet (*Vespula maculata*) lives throughout North America. This hornet is black with white marking, and is about 3 centimetres long. The community lives in a grey nest, which usually hangs from the limb of a tree.

The Yellow Jacket Hornet (*Vespula* species), named for its yellow markings, has its nest near, or even under, the ground. Up to 15,000 of these hornets can live in a single nest. They are quite aggressive, and there have been instances of many Yellow Jacket Hornets attacking a single person simultaneously. Some people are sensitive to the venom in their sting.

Two European species of hornets are now common in the USA. *Vespula germanica* entered the USA earlier this century. Another hornet, the Giant Hornet (*Vespa crabro*), arrived in the eastern states of the USA around the 1850s. This red–brown and yellow hornet grows up to 2.5 centimetres long. It will build its brown nest in crevices in rocks, inside hollow trees, or on buildings.

Hungry hornets

Autumn is a hard time of the year for the hornets. They need lots of protein to feed the hundreds of new babies (new queens and new males) being reared in the nest. So in autumn, many individual Giant Japanese Hornets will go out a-foraging for protein (which usually means the larvae of another insect).

The Giant Japanese Hornet (*Vespa mandarinia japonica*) has a well-organised routine that it follows when it goes looking for dinner. First, a single hornet will find the nest of the honeybee, kill a few individual bees and take them back to its nest to feed to its larvae. It will repeat this a few times. Then it will return, but this time it will actually mark the nest by rubbing a pheromone (a special attractant chemical) on, or near, the beehive. The Giant Japanese Hornet has a special gland (the van der Vecht gland) which makes this pheromone.

Once the beehive has been marked, the pheromone smell will attract other Giant Japanese Hornets. Soon another hornet will come to the beehive to join the first hornet, and the two of them will start hunting down bees individually. But once there are three or more hornets gathered together, their hunting behaviour will suddenly change. They will begin to start slaughtering the

DISTRIBUTED INTELLIGENCE

An individual insect has an IQ pretty close to zero, and can't do very much at all. But once a group of insects gather together, suddenly they show signs of intelligence and organisation.

In a nest, insects can organise air conditioning (by opening and closing flaps to control external wind currents, and by flapping their wings to make heat), nurseries for the babies (where special workers bring up the babies) and pantries (by going out and bringing in food, which then gets passed on to other workers, some of which sometimes even cultivate gardens).

Some Artificial Intelligence scientists are studying insects, to see how they do this. (Artificial Intelligence (AI) is the study of trying to copy human mental functions by computer programs.) Perhaps the individual insects just all follow a few local 'rules'. Perhaps the behaviour patterns of the individual insects then interact with each other, to give the impression of a bigger intelligence. Until recently, most AI scientists used a central intelligence in their simulations and in their robots. But now, many of them are looking at a 'distributed intelligence', which is what the nesting insects seem to use. So if AI scientists want to build a walking robot, perhaps they will install some tiny brains inside each foot, leg joint, and leg bone, rather than have the robot's movements all controlled from a single large brain.

honeybees *en masse*. One single hornet can, with its mandibles, kill a honeybee every 1.5 seconds. Allowing for little rests, a group of 25 or so hornets can kill an entire colony of 30,000 bees in just three hours. The hornets will then move in and occupy the nest for about ten days or so. During this time, they will busily carry the honeybee larvae and pupae back to their own hornet nest, to feed their own baby larvae.

One curious thing about the hornet–honeybee relationship is that the imported European Honeybee (*Apis mellifera*) has absolutely no defence against the hornets. It cannot recognise the special pheromone smell that the hornet uses to mark the nest. And it does not have an organised defence against the hornet. This probably shows that the Giant Japanese Hornet and the European Honeybee did not evolve side-by-side, in the same area.

But the native Japanese Honeybee (*Apis cerana japonica*) does have a well-organised defence against the Giant Japanese Hornet, which seems to imply that they evolved side-by-side, over a long period of time.

Not only can the Japanese Honeybees recognise the special pheromone smell, they can even recognise the peculiar movements that the hornet makes, as it comes in to mark the nest with the pheromone.

Honeybees' defence

This research into the defence system of Japanese Honeybees was started by Masato

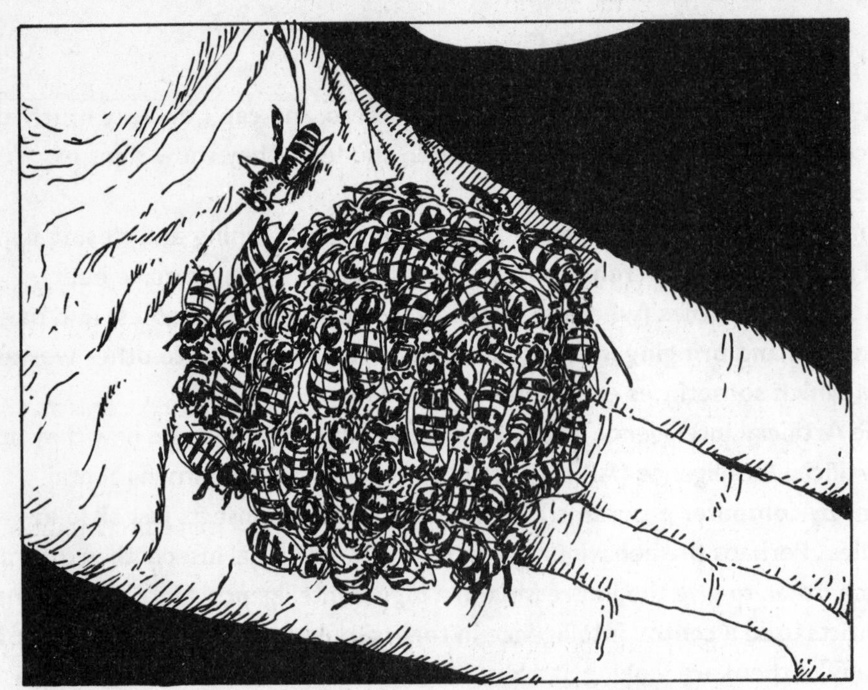

Ⓐ Defensive ball of APIS CERANA JAPONICA with ~ 400 tightly packed bees showing no stinging.

Ⓑ Predator killed by heat.

Ono and his colleagues from the Laboratory of Entomology at the Faculty of Agriculture, Tamagawa University, Tokyo, way back in 1984. They discovered the unusual thermal defence that Japanese Honeybees use against Giant Japanese Hornets. What aroused their curiosity in the first place was the discovery that the honeybees could somehow kill the hornet, but with absolutely no signs that the dead hornet had been stung!

The Japanese Honeybees have a very good game plan. They do not go out and attack the hornet marking their nest with the pheromone, but they deliberately provoke and annoy it. Over 100 worker honeybees gather together, close to the entrance of the nest. When the hornet comes near, they lift and shake their abdomens in a peculiar dance, which the hornet finds very aggravating. The bees then immediately dive inside their nest, as if taunting the hornet to follow.

In the same way that the hornet has a pheromone to call fellow hornets, it seems that the honeybees have their own special pheromone (isoamyl acetate) to recruit fellow honeybees.

The hornet doesn't know this, but over 1,000 worker bees have secretly gathered together, and they're waiting just inside the entrance. Once the hornet enters the nest, about 500 worker bees quickly engulf it, making a ball of bees roughly the size of a clenched fist.

The bees do not attempt to sting the hornet. Instead they cluster up as close as they can to the much larger hornet, and vibrate their muscles. This very quickly generates lots of heat.

Too high a temperature, for too long, will kill any living creature. It turns out that the Japanese Honeybee dies at temperatures over 49 Celsius degrees, but the Giant Japanese Hornet dies at a lower temperature — around 45 Celsius degrees. The honeybees rather cleverly split the difference. In less than five minutes, they get the temperature of the ball up to 47 Celsius degrees (above the lethal temperature for the hornet, but below the lethal temperature for the bees), and keep it there for about 20 minutes. Like good little soldiers, the bees have very good discipline, and do not try to escape from the ball, even though the hornet will kill a few bees before it dies.

Usually, the Japanese Honeybees are successful in defending their nest. However, when there are not many honeybees to defend the hive, it's the hornet that sometimes wins out.

But most times, the bees' efficient Neighbourhood Watch system wins the day, and the bees will roll the hornet, by shakin' all over, and makin' Great Balls of Fire!

REFERENCES

'Bees turn the heat on hornets', *New Scientist*, no. 1613, 19 May 1988, p. 35.

Evans, Howard E. & O'Neill, Kevin M., 'Beewolves', *Scientific American*, August 1991, pp. 56–62.

Exploring the Secrets of Nature, Reader's Digest Association, London, 1994, pp. 106–108.

'Great balls of fire', *Nature*, vol. 377, 28 September 1995, pp. vii, 289.

Mestel, Rosie, 'Hive from hell roasts hunting hornets', *New Scientist*, no. 1997, 30 September 1995, p. 18.

Ono, Masato, et al., 'Unusual thermal defence by a honeybee against mass attack by hornets', *Nature*, vol. 377, 28 September 1995, pp. 334–336.

SEX, SMELL & SEPARATION

What are you doing after Lab?

According to some recent research, women can 'sniff' out those men who have the genes that will give their baby a 'strong' immune system. It seems like a classic case of 'the nose knows'.

But unfortunately, it appears that the oral contraceptive pill (the Pill) actually reverses women's normal smell preferences (as if it wasn't already hard enough to find Mr Right!). In fact, it seems highly likely that a woman on the Pill will be attracted to the wrong man, if she follows her nose!

Humans are complex

We humans are very complicated animals, and that goes double for our emotions. Sometimes, it's almost impossible to work out why we like one person, but dislike another. People find different things attractive about other people — their looks, their physical and mental skills, or how much money and power they have. Sometimes, it seems that even another person's smell can subconsciously affect our feelings towards that person.

But how can a woman's sense of smell pick out a man who has the 'right' DNA to make a baby with a strong immune system?

One explanation reckons that it happens because there's a strong link between your

WHERE YOUR ODOURS COME FROM

There are three main sets of skin glands that can be involved in smells, or odours: the *sebaceous* glands, the *sweat* glands, and the *apocrine* glands.

The *sebaceous glands* are located most frequently around the scalp and face, genitals, and armpits. (There are none on the soles of the feet, nor on the palms of the hands.) They are usually next to a hair follicle. They dispense sebum, which is made up of fats that protect your skin, and protective chemicals that kill bacteria. The fats are composed of triglyceride fats (about 60 per cent), wax esters (about 25 per cent), and the remainder squalene (12 per cent), cholesterol esters (about 3 per cent) and cholesterol (1.5 per cent).

Some bacteria are not stopped by the sebum's protective chemicals. So these bacteria split the fats into smaller chemicals, which can have noticeable odours.

About one-third of the weight of the fatty acids on the skin is common fatty acids that come from the sebaceous glands (fatty acids such as palmitic acid, linoleic acid and oleic acid). But the remaining two-thirds are made up of several hundred unusual fatty acids that come via the activity of bacteria.

The sebaceous glands grow rapidly during puberty. On the genitals, the sebaceous glands make smegma.

The *sweat glands* discharge water (about 99 per cent) and salt and amino acids (about 1 per cent). They open directly onto the skin. In most healthy people, they are usually not involved in generating body odours. In your armpits, your sweat glands are closely linked to your apocrine glands, which do generate odours.

The main job of sweat is to cool you down, via evaporation. On the soles of the feet and the palms, sweat actually improves your grip (unless there is a huge amount of sweat). Sweat also keeps the skin on your soles and palms soft, so that they can be as sensitive to touch as possible.

Men and women have the same numbers of sweat glands: about 150 to 350 in each square centimetre, making a total of some 2 million to 5 million over the whole body. The greatest density of sweat glands is on the palms and soles, followed by the head, trunk, and the limbs. Some sweat glands discharge sweat when they're stimulated by heat, while others discharge sweat in response to psychological stimuli (such as when you're scared).

An *apocrine gland* is usually associated with a hair follicle, and opens onto the skin close to that hair follicle. Even though apocrine glands are fairly large, they discharge only small quantities of a viscous, milky fatty liquid, which can be off-white, yellow or red in colour. Bacteria usually break down the fats into smelly, free fatty acids. Many deodorants work by killing these bacteria. These glands develop rapidly during puberty.

Apocrine glands are mostly found in the armpits and around the genitals, but they are also found on the scalp, hands, cheeks, ears, and around the nipples.

APOCRINE GLANDS, EVOLUTION AND COMMUNICATION

When you were still in the uterus, about five-and-a-quarter months after conception, apocrine glands began to sprout all over your body. But within a few weeks, most of them disappeared. This is very odd.

Some scientists have used this information about the sprouting, and withering, of apocrine glands to come up with a theory. They speculate that perhaps our non-human ancestors had apocrine glands all over their bodies, and that as we evolved into humans, we gradually lost them.

Certainly, we can see this pattern in the two-legged mammals, the primates. We humans have hardly any apocrine glands and lots of sweat glands, the great apes have roughly equal numbers of each, while other (lower) primates have more apocrine glands than sweat glands.

Another strange thing about apocrine glands is that we don't seem to consciously smell their subtle odours. In all the other animals, apocrine glands emit odours which the animals use for 'telling' each other about their position in society, and for trying to attract members of the opposite sex. We humans can register that there is a smell coming from the armpit of another person. But we can't interpret this smell to see if the other person is (for example) sexually available or not. But any animal can tell this immediately from the smell.

natural body odour, your sense of smell, and your immune system.

The task of your immune system is to recognise, and then fight off, invaders. So one job of the 'recognition sub-system' of your immune system is to pick up incoming foreign chemicals. These invaders can range from spider poison, to toxins in spoiled food, to the smell of another person.

This surveillance duty is controlled by the Major Histocompatibility Complex (MHC), which is a part of our DNA. The MHC controls the 'eyes' of the immune system. The 'teeth' that actually destroy or immobilise the invaders are controlled by another part of the DNA.

When things go wrong with the MHC, you can suffer from the so-called 'auto-immune' diseases — diseases in which your immune system attacks *you*, instead of attacking invaders. These diseases include ankylosing spondylitis, ulcerative colitis, Hodgkin's Disease, psoriasis vulgaris, juvenile diabetes mellitus, myasthenia gravis, polycystic kidney disease and paralytic polio.

The MHC has another function — it helps control your body odour. In fact, some of the odorous chemicals that leave your body via your various glands are actually breakdown products of chemicals of the immune system. So there's a strong link between your body odour and your immune system.

A rodent knows

We humans don't seem to be strongly affected by odours. Maybe we have a lot of social conditioning. But some other mammals are extremely sensitive to smells.

Consider these three experiments.

First, the smell of an unfamiliar male mouse can cause an abortion in a pregnant female mouse! Suppose that a female mouse is pregnant to a certain male mouse. Suppose that the odour of a new male mouse (one that the female mouse has never before met) is wafted over the newly pregnant female mouse. Suddenly, the normal hormonal changes of pregnancy stop happening in our female mouse, and the fertilised egg is aborted! This is called 'pregnancy block'. This experiment links the sense of smell, odours, and fertility. (An opposite effect also exists. A chemical in the urine of a male mouse can start up, and even speed up, the fertile cycle of a female mouse.)

Second, consider the well-known laboratory experiment of rats and mazes. It turns out that if they are given a choice, male rats will prefer to run in a maze that has the smell of a sexually available female rat.

AMAZINGLY SENSITIVE NOSES

According to the 1993 *Guinness Book of Records*, the most acute sense of smell belongs to the male emperor moth *(Eudia pavonia)*. Back in 1961, German scientists found that the male can pick up the odour of the virgin female emperor moth at a distance of about 11 kilometres, upwind! The sex attractant chemical is an alcohol ($C_{16}H_{29}OH$). This chemical is so effective on the male moths, that the females don't need to carry much — less than 0.0001 milligram!

The female silkworm moth (*Bombyx*) uses bombykol as its sex attractant. Bombykol is so powerful, that the amount that it carries (1.5 micrograms), is enough to attract 1 billion males!

In general, we humans are not noted for having a very sensitive sense of smell. But we are exquisitely sensitive to ethyl mercaptan — the chemical that makes rotting meat so pungent. Your nose can detect ethyl mercaptan even when it's diluted to as little as 1/400,000,000th of a milligram in each litre of air!

The sense of smell becomes more sensitive when we are in pain, such as might be caused by a pinch on the foot. This is a protective reflex.

The sense of smell of women becomes more sensitive around ovulation, especially to odours that are related to sex hormones.

HOW WE SMELL ODOURS

There are two areas in our nose that we use to detect odours — the well-known 'olfactory epithelium' and the not-so-well-known 'vomeronasal organ' (VNO).

We also detect odours by means of the pain endings of the trigeminal nerve. These pain endings are widely scattered over most of the inside of the nose, and they respond to major irritants such as sulphuric acid, and minor irritants such as vinegar.

Physiologists have known about the olfactory epithelium for a long time. 'Olfactory' is a technical word meaning 'related to smell'. The olfactory epithelium is a patch of yellowish tissue high up in the roof of the nose. Normally, it is poorly ventilated with air, but when we sniff deeply, we pass lots of air over it.

In this yellow patch, there are sensory cells specially adapted for smelling, as well as supporting cells to hold the structure together. There are also gland cells that supply mucus to cover this yellow patch. Chemicals in the air enter the nose, dissolve in this wet layer, and then excite the sensory cells.

Your human yellow patch has about 40,000 sensory cells per square millimetre, with a total area of about 250 square millimetres, giving a total of about 10 million cells that can detect odours. (The rabbit has about 120,000 per square millimetre, giving a total of 100 million sensory cells — with a total surface area greater than the skin of the rabbit's body!)

There is also another area in the human nose that we can detect odours with — and until recently most scientists didn't believe it existed! This is the VNO, also known as the Jacobson's Organ in other animals.

Snakes use it, combined with their forked tongue, to give themselves 'stereo smell-o-vision'. Fishes, birds, and some mammals don't have a VNO, but it is very well developed in snakes and lizards. In snakes, it's located in the roof of the mouth, but in humans it's near the wall that separates the nostrils, about 2 to 3 centimetres inside the nose. There's one VNO in each nostril.

But back in the 1930s, physiologists said that not only did we humans definitely not have a VNO, there was no structure in the brain to process the information from any such organ. However, since the early 1990s, Luis Monti-Bloch and his team from the University of Utah have used modern microscope technology to find the human VNO. We also now know that nerves lead from the human VNO to the accessory olfactory lobe of the brain.

Most people don't consciously realise when their VNO is stimulated. We can waft a smell up their nose, and their VNO can fire frantically with electrical activity — but all they get is a vague, generalised emotion of feeling fine.

So, while the olfactory epithelium seems to detect those smells that we consciously register, the VNO picks up the smells that have a deeper, and more subtle, effect.

Perhaps the odours that subconsciously affect our behaviour come from our apocrine glands, and are detected by our VNOs.

Thirdly, consider a cute experiment that was done with mice, way back in 1976. This experiment tied together sexual preferences, odours and the immune system.

A male mouse was given the choice of mating with one of two female mice. One female mouse gave off an odour that was similar to the odour of the male mouse. Her MHC (which controls part of the immune system) was very similar to the MHC of the male mouse that she smelt like. The other female gave off a different odour, and had a dissimilar MHC.

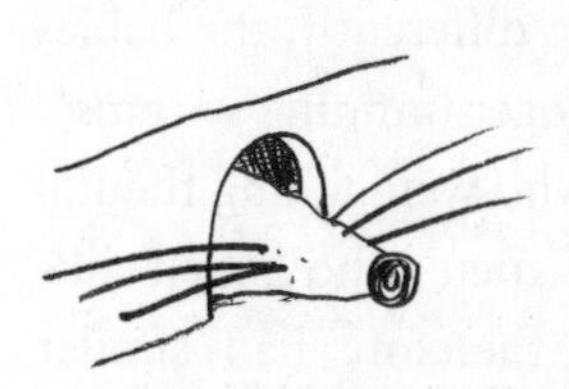

The male mouse preferred to mate with the female mouse that smelt different from him, and therefore had a different MHC and different immune system!

The advantage of mating two mice that had different immune systems was that their baby mice would have a wider range of immune responses. This would give the babies better chances of survival in later life.

Suppose that the daddy mouse was resistant to one group of diseases — call them A, B and C. Suppose also that the mummy mouse was resistant to a different group of diseases — call them D, E and F. Then the baby mice would be resistant to both groups of diseases — A, B and C, and D, E and F.

But if two similar mice mated, then the babies would be resistant only to one group of diseases.

Over the years, there have been a few 'odd' findings about human fertility and MHC. These all compare couples with similar MHCs, to couples with dissimilar MHCs.

If a couple who are trying to 'get pregnant' have similar MHCs, they will have more difficulty conceiving a child. Couples with similar MHCs will also have more spontaneous abortions. And finally, babies of couples with similar MHCs will tend to be slightly underweight.

These are some of the reasons for believing that there are advantages in the potential parents having different MHCs, and different immune responses. Certainly, the experiment with mice showed that

KISS OR SMELL

We do have one custom that gives us a chance to smell another person — the social kiss.

Eskimos and Polynesians don't just rub noses — they also take deep smells of each other. There is a similar custom among the Sami of Europe, and in southeastern India, where one person presses his or her nose up against the cheek of the other person, and then inhales deeply.

Early Christians used the kiss as a greeting. This habit still occurs in Roman Catholic ordinations and consecrations. Even today, we still have the habit of kissing the bride. But in most of Western society, we have passed up the chance to use a kiss as a chance to smell.

Maybe the 'air kiss' could perform some useful social function after all, and give us a chance to smell our friends!

male mice preferred to mate with females who smelt different, and who therefore had different immune systems.

T-shirt sniffers

Claus Wedekind, from the University of Bern in Switzerland, recently did a similar experiment to the mouse-smell experiment of 1976 — but with 100 humans!

Now it's hard to imagine how you would word your application to your local Ethics Committee, if you wanted to run an experiment where you put different-smelling people in different cages, and watch to see with whom they mate. So Wedekind made a compromise.

He just asked his volunteer men to sleep in the same T-shirts for two nights. He wanted the T-shirts to be drenched in the men's natural odours, and nothing else. He also asked the men not to use any products that had a smell (such as underarm deodorants, perfumed soaps or cologne), not to smoke or drink, not to eat spicy foods, not to sleep next to another person, and not to have sex. And yes, he did have a bit of trouble getting enough volunteers.

The precious sweat-soaked T-shirts were stored in a plastic bag during the day.

Then, Wedekind and his team took the sweat-soaked T-shirts to his volunteer women. One at a time, a female volunteer was put into a room all by herself. First, she was asked to smell a brand-new T-shirt that had never been worn (this was to check that the T-shirt itself was neither incredibly attractive, nor repulsive). Then she was asked to smell the T-shirts, one at a time. Finally, she was asked to rate the T-shirts on a scale that ran from very attractive to very repulsive.

Now, some of the women were taking the Pill, and some were not. Both the men and the women had blood tests taken at the start of the experiment, to see what kind of MHC they had.

The women who were *not* taking the Pill had results just like the mice experiment of 1976. They were attracted to the smell of the T-shirts that belonged to men who had a different smell, different MHC, and different immune system. The more different the MHC, the more they liked the smell. Presumably, if the women had babies with men who smelt differently, the babies would have had stronger immune systems.

But the women who were taking the Pill liked the smell of men who smelt like themselves — and therefore had similar immune systems! As Claus Wedekind writes: 'The contraceptive pill seems to interfere with natural mate choice.'

Pregnancy and the Pill change smell

Now, the Pill mimics the effect of being pregnant. So here's a half-way believable explanation, in terms of our distant past.

Long ago, before we had invented digital watches, a woman would go out to find a man who smelt differently, and so had a different immune system. They would mate, and she would become pregnant — pregnant with a baby who would have a 'strong' immune system. This baby would have a better chance of surviving than a baby with a 'weak' immune system. Then, with the hormones of pregnancy racing through her body, her sense of smell would change, and she would prefer the company of people who smelt like her. So

SMELL AND BLOODHOUNDS

It has been known for a long time that trained bloodhounds are very good at tracking down humans by their smell alone. But they cannot pick the difference between identical twins!

This is probably because identical twins have identical MHCs, and identical smells.

On the other hand, if you give them the smell of one twin, they can track down the other twin!

she would go back to these people who smelt like her (her own family), who would take care of her while the baby grew inside her.

But in our modern society, when a woman goes out to find a mate (who might become the father of her future children), she's often already on the Pill. As the Pill mimics the effect of being pregnant, she'll prefer the smell of a man who has a smell like her own.

But if she stops taking the Pill with the plan of getting pregnant, she'll suddenly find that she doesn't like the smell of her mate any more!

If she's lucky, she might manage to ditch the bad-smelling man in her bed before they make a baby. Otherwise, she might have to look forward to a messy divorce.

In Western society, there has been a big increase in the divorce rate ever since the Pill was introduced. (At the same time, however, there has also been a changing of the divorce laws, and people's attitudes have become more relaxed.)

Letting your heart follow your nose

It's probably a bit impractical to suggest that people go around sniffing their prospective partners, and choosing a mate is much more complicated than subjecting a person to a sniff test.

First, we humans have had many layers and many years of social conditioning. We very rarely follow our basic instincts.

Second, most of us are not happy with our natural smell, for many different reasons. Even if we don't wear perfume or cologne, most of us wash with a soap that is slightly perfumed.

Maybe the resurgence of the condom will mean that women can now let their hearts follow their noses to their dream date.

And maybe your parents were right after all — maybe opposites do attract!

REFERENCES

Aldhous, Peter, 'Darling, you smell wonderfully different', *New Scientist*, no. 1976, 6 May 1995, p. 19.

Furlow, F. Bryant, 'The smell of love', *Psychology Today*, March–April 1996, pp. 6, 38–45.

'The noses have it', *Scientific American*, November 1995, p. 14.

Wedekind, Claus, et al., 'MHC-dependent mate preferences in humans', *Proceedings of the Royal Society of London*, vol. 260, 1995, pp. 245–249.

SILK STOPS BULLETS

'A somewhat extensive experience in the gunshot wounds of civilian life, during the past few years, has brought to my attention … instances illustrative of the remarkable tenacity of silk fibre and its resistance to the penetrative power of a bullet' — George Emery Goodfellow, surgeon, Tombstone, Arizona, 1887

Silk yarn

Way back in 1887, an American medical doctor with the unlikely name of George Emery Goodfellow wrote an article for a medical journal called the *Southern California Practitioner*. He called the article 'Notes on the Impenetrability of Silk to Bullets'. In this article, he described three gunfights, which he had personally witnessed at close range, in which ordinary silk had stopped bullets! The silk had been woven into thin handkerchiefs or scarves — so less than a millimetre of silk had stopped a bullet!

Doctor Goodfellow lived in the Wild West town of Tombstone, in Arizona. He was a skilled surgeon, and quite an expert on gunshot wounds.

Writing about one case, he says: 'In the spring of 1881 I was a few feet distant from a couple of individuals who were quarrelling. They began shooting.' Doctor Goodfellow goes on to describe how one of the gunfighters received two bullets in his chest, and died. He continues: 'The ball came from a cut-off Colt 45-calibre revolver, fired at a distance of six feet, the cartridge of which contains thirty grains of powder and two hundred and sixty grains of lead … Not a drop of blood came from either of the two wounds. From the wound in the breast, a silk handkerchief protruded.'

When Doctor Goodfellow pulled on the ends of the silk handkerchief that protruded from the entrance wound, the handkerchief came out — and it brought the bullet with it! Even though the bullet was caught by the silk, the man had died, because the bullet had penetrated his chest cavity. The bullet had easily smashed its way through the man's clothing and body tissues — but it could not get through two thicknesses of a silk handkerchief that the man had worn in a pocket!

Goodfellow later fired an identical bullet with an identical charge of powder, from an identical gun — and found that the bullet could smash through 10 centimetres of pine timber! And yet, a millimetre of silk could stop what 10 centimetres of wood could not.

In another similar case, he described how a silk scarf, which was wrapped around a man's neck, also stopped a bullet. 'The life of this man was, presumably, saved by the handkerchief,' he wrote.

These days, this strange property of a fabric to stop a bullet is used in 'Fashion-conscious ballistic apparel'.

'Fashion-conscious' is a phrase meaning that it comes from a small-volume designer, and so there might be a groovy label worn on the outside for all to see and admire. 'Ballistic' means that it's related to moving objects, such as bullets or projectiles. 'Apparel' means that it's some kind of fabric, that you can wear. In other words it's soft body armour, or bullet-proof clothing.

The history of soft body armour goes back a long way. The Greeks wore leather armour to protect themselves from spears and arrows. In World War I, British pilots discovered that thick layers of silk stopped low velocity shrapnel better than steel. They wound the silk around their heads, and then wore leather horse-riding helmets on top of the silk. In World War II, nylon was used for flak jackets. Nowadays, a synthetic material called Kevlar is the popular choice.

In a few years, synthetic silks might appear in parachutes and seat belts, and hard-wearing clothes. They might even appear in solid objects such as body panels and bumper bars on cars.

So the next time the posse comes looking for you, maybe you should head straight for the lingerie drawer!

Stronger than steel

Many creatures make a silk, but so far, only the larvae of the moth *Bombyx mori* makes a silk that can be spun commercially.

Silkworms gorge themselves on the leaves of the white mulberry tree. Over a period of just four weeks, they will increase their weight by a factor of 10,000 times! They will shed their skins four times, before they finally begin to spin their cocoon of silk. Once the cocoon is spun, they are killed — usually with hot water or steam. (If they were allowed to live, they would break their way out of the cocoon and spoil it.)

The hot water or steam also weakens the

AERODYNAMICS OF

Catching a fast-moving insect in a web is like catching a fighter plane with a net that has ropes 25 millimetres thick!

Insects are small, but they can move rapidly, so why doesn't the web break, and let the insect through? Well, thanks to high-speed video cameras, and a computer program used to simulate how cars crumple in a crash, some English zoologists and physicists have found that the wind resistance of the spider web plays a big part in catching dinner.

The orb web is a pretty neat two-part design.

First, it has stiff, dry threads that radiate out from a common centre — a bit like the spokes on a bicycle wheel. These strands provide a stiff, slightly stretchy framework for the other type of thread.

Second, the web has a stretchy, wet, sticky spiral thread that winds out in increasing spirals from close to the centre. This wet spiral thread is the capture thread. It's very stretchy. This is very handy for absorbing the energy of a speeding insect.

However, it's not quite so simple. Spider thread is very thin — about one-millionth of a metre in diameter, or one micron. The incoming insect is moving at a few million microns per second. If you do the physics, and work out how much energy is absorbed in stretching the web, you find that so much energy is left over that the web should break. But the web doesn't break! So where is the extra energy in the incoming insect going to?

Lorraine Lin, Donald Edmonds and Fritz Vollrath from the University of Oxford discovered this aerodynamic secret. They used high-speed video cameras to see what happened when insects crashed into the web of the Garden-cross Orb Spider (*Araneus diadematus*). They also used a non-linear structural analysis finite element computer program, which they borrowed from the car industry. This program, called OASYS DYNA–3D, was made to test the 'crash worthiness' of cars, but it also works just fine on spider webs. Their theories were probably correct, because they saw the

sticky protein gum that binds the silk together. Then a human, with a lot of patience and fine motor skills, has to find the free end of the silk thread, and unwind it. The silk from a single silkworm cocoon can be up to 1.6 kilometres long!

Silkworm silk is a magnificent material. It's lighter than cotton, thinner than a human hair, and is three to five times (weight for weight) stronger than steel. It's strong, it's elastic, and it's tough. Even though we have now made other materials that have similar properties, it seems that Nature, with a few hundred million years of evolution up her sleeve, still has the superior products.

THE SPIDER WEB

same things on the video and on the computer screen!

They saw that after an insect impact, the movement of the web was confined to within three radial spokes either side of the impact site. In other words, the effect of the impact was limited to within six radial spokes, and the rest of the web was unaffected. These radial spokes or threads provided most of the forces that brought the web back to its original shape. They also saw that the movement of the web very rapidly died away — it didn't oscillate back and forth for a long time.

But usually, whenever anything is pushed, there is a tendency to return to its original shape. This bouncing-back like a trampoline could shoot the insect free, back in the direction from which it came.

This is where the spider's web uses some very clever aerodynamics.

The wind resistance of the size of objects that we normally deal with (like a hand held foolishly out of a car window) depends on the size of the object. But the wind resistance of something as skinny as a spider's web has nothing to do with the size of the thread! The wind resistance depends only on how fast the web is moving — so the faster it moves, the greater the wind resistance.

When a fast-moving insect hits the web, it tries to move the web rapidly. This automatically creates wind resistance, which tries to slow the web down. So the web stops moving really rapidly.

But the down side of this wind resistance also means that the web can be easily destroyed by a strong wind. This is bad news, because it means that the spider has to make a new web from scratch, instead of recycling the old web (by eating it, and then extruding another one).

But the spiders have been making circular orb webs for 180 million years, which is about 60 times longer than we humans have been around. This probably means that they've accurately worked out all the cost–benefit ratios of the skinny web — even without computers or video playback!

Crystal block and spiral proteins

But how does something soft, like silk from a silkworm, stop a bullet? We don't really know yet, but scientists in the military and in the universities are looking at silk from silkworms and spiders to try to find out why.

Part of the secret lies in proteins. In silk there are two main types of proteins: crystal blocks and spirals.

First, let's look at the blocks of crystal. Some of the proteins in silk are folded in pleated sheets, a bit like a pleated netball skirt. These sheets are neatly stacked up on top of each other, like a piano accordion, to

MAD SPIDERS, DRUGS AND WEBS

If the blood of a schizophrenic person is given to a spider, it will weave very strange webs. Gradually with time, as the blood is fully broken down, the spider will start to weave more normal webs. (Just for comparison, the blood of a non-schizophrenic, otherwise healthy, person will have no effect on a spider's web.)

NASA scientists at the Marshall Flight Centre in Alabama found that different drugs will change the web that a spider will spin.

Speed (amphetamines) will set the spiders working furiously, but in an undisciplined fashion — so the spiders leave huge holes in the webs. Caffeine makes them lose their way entirely — the webs are just a few threads that are randomly strung together. Chloral hydrate makes them so sleepy that they can hardly even start work. Marijuana lets them get started, but then they stop — presumably, the spiders get so legless that all they can do is admire their handiwork!

make a large crystal. Bunches of these highly organised crystal blocks are embedded in a mushy sea of tangled proteins, which are a bit like spaghetti.

The second kind of protein is this spaghetti-like protein. But whereas ordinary spaghetti is quite smooth, these proteins in the sea of mush are all spirals, like corkscrews. Because these spiral proteins are all very randomly and chaotically entangled with each other, it takes a lot of force to pull them apart.

So a silk thread is made from two main elements: blocks of tightly folded proteins in a sea of tangled spiral proteins. It's very elastic, because each of the elements (the blocks and the spirals) is individually very stretchy. It's very strong, because it's a composite, like fibreglass (which has fibres of glass embedded in a resin).

A spider's superior silk

Silkworms have been making their silk for us for centuries. But spiders make silk as well.

There's a good reason for to trying get hold of spider-web silk — and that's because spider web silk is magical stuff.

In fact, a spider doesn't make just one type of silk. One spider can make many different types of silk, for many different uses. It can make a tough silk for the outer part of the egg sac, and a softer silk for the inside. It can make an elastic silk for the core fibres of the capture spiral, and it can make a more flexible silk for the drag line.

Recent research has shown that there are consistent differences between these silks, in their amino acid make-up. But this is just one factor. Another factor is how the silks are extruded as threads from the spider's body.

The Golden Orb Weaver (*Nephila clavipes*) is a large non-poisonous spider. Its

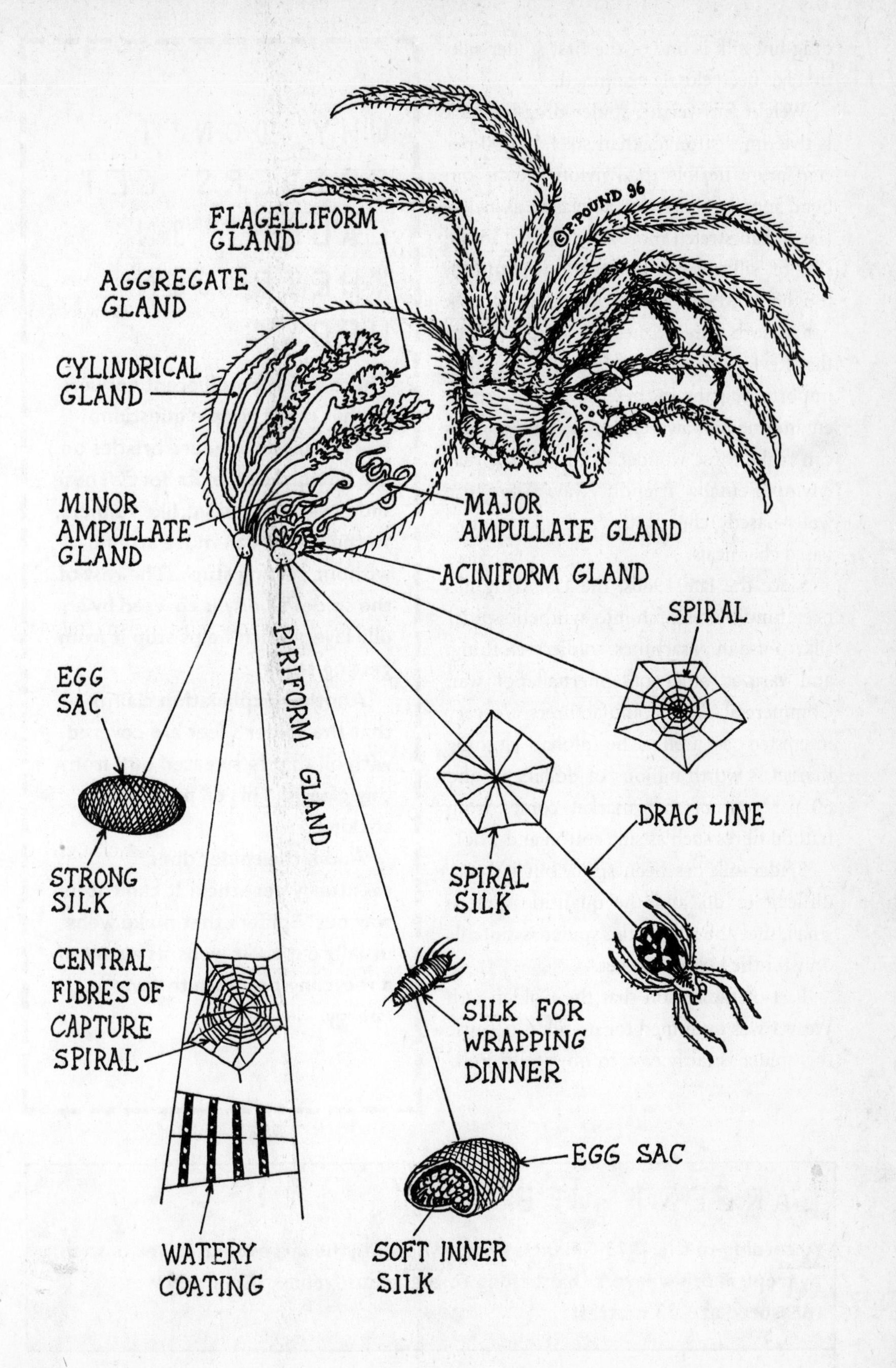

©P.POUND 96
FLAGELLIFORM GLAND
AGGREGATE GLAND
CYLINDRICAL GLAND
MINOR AMPULLATE GLAND
MAJOR AMPULLATE GLAND
ACINIFORM GLAND
PIRIFORM GLAND
SPIRAL
EGG SAC
STRONG SILK
DRAG LINE
SPIRAL SILK
CENTRAL FIBRES OF CAPTURE SPIRAL
SILK FOR WRAPPING DINNER
EGG SAC
WATERY COATING
SOFT INNER SILK

drag-line silk is one of the first spider silks that has been closely examined.

Weight for weight, spider drag-line silk is five times stronger than steel, but 30 per cent more flexible than nylon (i.e., it can bend more), and twice as elastic as nylon (i.e., it can stretch more). Compared to the silk of silkworms, it's more waterproof, tougher, and can stretch more. Spider silk can absorb three times the impact force that Kevlar can, without breaking. Another important factor, in these days of increased environmental awareness, is that spiders can make these wonderful fibres in a very environmentally friendly way. They use water-based chemicals, not petroleum-based chemicals.

Since the late 1960s, the US Army has been funding research into synthetic spider silk, to use in parachutes, soldiers' clothing, and various other paraphernalia of war. Commercial cloth manufacturers are very interested as well. The global clothing market is worth billions of dollars — and 60 per cent of that market comes from natural fibres such as silk, cotton and wool.

Spider silk has been spun, but it's very difficult to do, and the quantities are so small, that the silk of the spider is suitable only for the luxury market.

Part of the reason that the Golden Orb Weaver was examined for its silk is because this spider is fairly easy to physically 'milk'

WHY DON'T SPIDERS GET CAUGHT IN THEIR OWN WEBS?

There are a few different answers to that really simple question.

The spider has hard bristles on the bottom side of its 'feet'. These slide over the thread, like runners, so the spider can move around without getting stuck. The rest of the spider's body is covered by an oily layer, which helps stop it from getting stuck.

Another explanation claims that the spider's feet are covered with oil that is 'sweated out' from tiny glands. This oil prevents sticking.

And if the spider does eventually get stuck? It can eat its way out! Spiders that make webs usually eat their webs, as a part of a recycling program to save energy.

LARGEST WEB

According to the 1993 *Guinness Book of Records*, the largest aerial web is spun by tropical orb weavers that belong to the genus *Nephila*. One web was measured at 5.73 metres!

EATING THE WEB

It takes metabolic energy to make the proteins that are in a web. This is part of the reason that spiders eat their own webs, as part of an energy-efficient recycling program.

But another reason for eating one's web might be to get water! The Garden-cross Spider (*Araneus diadematus*) usually eats its web just before dawn each day, when the web is most saturated with water. This would give the spider 10 per cent of its daily water requirements. It would use this water to keep its lungs moist.

the silk from. But this 'milking' method would take forever to get enough silk just to make a pair of socks.

Cultivated spider silk

Back in the 18th century, French scientists decided that if the Chinese could cultivate silkworms, then they could cultivate spiders. Silkworms do make a very good silk, but the Chinese had enjoyed a virtual monopoly on silkworm silk for centuries.

So the French scientists tried to get the special silk made by spiders. To get enough spider silk to make just a single dress, they needed thousands of spiders. So they packed their thousands of spiders in a barn to collect the silk from their webs. But while silkworms are very calm and docile, spiders are aggressive, and a whole lot more territorial. The spiders spent their time fighting and eating each other, and not making silk. Soon, there were virtually no spiders left (and no silk either), so the French scientists had to abandon that particular method of getting spider silk.

Randolf Lewis, a molecular biologist at

DIFFERENT WEBS FOR DIFFERENT DINNERS ...

For a long time, it was thought that any particular species of spider would make only one type of web for catching dinner. But Christina Sandoval, now at the University of California at Santa Barbara, has found one spider that makes different webs for different dinners!

The spider is the South American *Parawixia bistriata*. It usually spins a fine-mesh web, around sunset each day. But Christina Sandoval noticed that during September, at different times during the daylight hours, the spider would spin a wide-mesh web. The web was the perfect size for catching flying termites, who would come out of their nests during the daytime and swarm, before going off to set up new nests!

MUMMY EATS DADDY

The Australian female Redback Spider (*Latrodectus hasselti*) has a really unusual mating habit. In 65 per cent of her sexual encounters, she will eat her male partner. And he goes willingly!

As soon as he has inserted his (delicately named) intromittent organ into her, he flips a somersault, so that his juicy abdomen is right in front of her mouth. He's not very big — only about 1 per cent of her body weight. But it's still enough of a juicy morsel that she starts chomping into him immediately. Despite this obstacle, what's left of him continues having sex, until there's not enough left of him to run his intromittent organ. Why?

For the male redback there are two advantages in being eaten alive during sex, according to Maydianne Andrade, a graduate student from the University of Toronto. First, he gets to have sex for longer, so he can deposit more sperm, so he has a bigger chance of passing on his genes. Second, if she is filled with sperm, she is more likely to reject the sexual advances of another male redback, again increasing his chance of passing on his genes.

There are a few practical reasons for going out with a bang. For one thing, male redbacks have a short life span anyway. Even if he doesn't get eaten during sex, the tip of his intromittent organ will break off during sex, so he can never use it again. As if that isn't bad enough, he will usually die very shortly after sex, anyway. It's probably better to go out on a high note.

BABIES EAT MUMMY

All mother spiders will give some degree of nurture and care to their baby spiderlings. This ranges from weaving a strong protective silk around the eggs, to guarding the babies after they are born and getting food for them.

But the Australian Social Spider (*Diaea ergandros*) Evans (Thomisidae) goes even further — she lets her baby spiderlings eat her alive!

These caring mother spiders start off by putting a protective layer of eucalyptus leaves around the spiderlings' nest, and then go off to capture and bring back insects. The spiderlings are relatively helpless in their early days, and need lots of care.

Soon afterwards, the ungrateful spiderlings start drinking a liquid (haemolymph) from the leg joints of their mother. She does not resist. Within a few weeks, she cannot move at all, and so the babies eat her alive, entirely!

It's better for the spiderlings if the mother is big and heavy. Light mothers are cannibalised early on, and fewer of the spiderlings survive. Who can say what's better for the mother?

the University of Wyoming, has found an easier way — he's using genetic engineering. He and his colleagues have uncovered two of the genes in the spider's DNA that make the drag-line silk. Drag-line silk is what the spider uses to abseil down from its secret hiding spots, or to make a frame for the perimeter and spokes of its web. So this particular type of silk has to be strong and elastic. The molecular biologists inserted these genes into a bacterium called *E. coli*. They now have vats full of mutated *E. coli* bacteria that will make proteins that are the same as those in spider web drag-line silk.

However, these proteins are inside bacteria, not neatly organised into a thread. Lewis and his colleagues have found a way to turn these proteins into a thread. They collected the protein, purified it, dried it and then re-dissolved it. Finally, they squirted it through a very fine hole. As the proteins were pushed through the tiny opening, the friction made the proteins get longer, and stick together. That's how they made a synthetic silk fibre, from natural proteins. The proteins are not as well organised as they are in the natural thread, but they are still pretty good.

Spider molecules

Spider drag-line silk seems to use different molecules from silkworm silk. Part of the reason that spider drag-line silk is so strong, yet flexible and elastic, is because it has a system of molecules that interlock. These molecules are like a strange combination of Lego and Slinky; Lego being the kid's toy with hundreds of tiny pieces that interlock with each other, and Slinky the coil that you can 'walk' down the stairs.

These spider molecules have different shapes and properties in the middle, and at the ends. In the middle, the molecules have a shape like a spiral spring — this Slinky-like shape makes the molecule elastic, and stretchy. But each end of this molecule is either concave or convex, just like Lego, so one molecule can 'snap' into another. This makes a very strong, and very flexible, connection.

For cables, tendons and knife-proof vests

If you could get enough of this drag-line silk, it would be really useful. You could make very thick cables, and hold up bridges with it. Thinner cables could catch landing planes on aircraft carriers.

You could make incredibly thin threads, and use them as stitches in operations, where they would cause much less scarring and bleeding than normal stitches. In fact, in folk medicine, spider silk has long been used to hold wounds shut. According to Lanny Johnson, an orthopaedic surgeon at Michigan State University, you could use spider silk in artificial tendons and ligaments. Our current artificial tendons and ligaments are made of synthetic materials, which lose their shape after a few years. But a biological material, like spider-web silk, should last longer.

Delwar Hussain, from the University of Wyoming School of Pharmacy, said: 'Spider silk is very resistant to climate changes, bacteria, enzymes and fungal growth. We think it could be a good substitute for sutures in tendon and ligament wounds, or with artificial prostheses.' He has used spider silk with his mice. His research shows that not only does spider silk remain strong and stable when used inside the skin and inside muscles, but also that it's not toxic to mice cells.

Unfortunately, your average household spider will spin only 2–3 milligrams of silk each day. You need about 5–6 milligrams for an average stitch.

You would need to use different silks for different jobs. For example, you couldn't use drag-line silk for a bullet-proof vest because it's too flexible for this job. Sure, it would stop the bullet, but the bullet would carry the vest with it and actually penetrate into your body before it finally stopped. For a bullet-proof vest, you need a different type of silk that would be stiffer and less flexible.

Another line of research is trying to make knife-proof vests. A knife moves much more slowly than a bullet, but the point is much smaller and concentrates the energy into a much smaller area of contact. So an ordinary knife can have three times more penetrating power than a bullet! Ian Horsfall and Celia Watson, materials scientists from the Royal Military College at Cranfield University in Shrivenham, are working on this problem. A complex tangled weave, made from very tough fibres, could be the solution.

There are some 30,000 to 40,000 different species of spiders on our planet. But only 10,000 of these species make a web to catch their dinner. (It takes a lot of energy to make a web, they are easily broken by birds and high winds, and the process of building an aerial web puts the spider out in full view of hungry birds.) Each of these 10,000 spiders can make half-a-dozen or so different silks.

So far, the molecular biologists have looked at only one of the silks, made by one of these spiders. This spider was chosen simply because it was easy to milk, not because its web had the best properties. Imagine what properties we will find after we check out the entire 10,000 species of spiders that make dinner-catching webs.

Then we could all be wearing bullet-proof Spider Man suits!

REFERENCES

Goodfellow, G.E., 'Notes on the impenetrability of silk to bullets', *Southern California Practitioner*, vol. II, no. 95, March 1887.

Graham, David, 'Synthetic spider silk', *Technology Review*, 16 October 1994, pp. 16–17.

Harp, Joel M., 'Bullets and silk in the Old West', *Science*, vol. 271, 2 February 1996, pp. 580, 581.

Hawkes, Nigel, 'Spiders on speed spin out of control', *Weekend Australian*, 29–30 April 1995, p. 16.

Lin, Lorraine H., et al., 'Structural engineering of an orb-spider's web', *Nature*, vol. 373, 12 January 1995, pp. 146–148.

Vollrath, Fritz, 'Spider web and silks', *Scientific American*, March 1992, pp. 52–58.

PLANTS MAKE PLASTIC

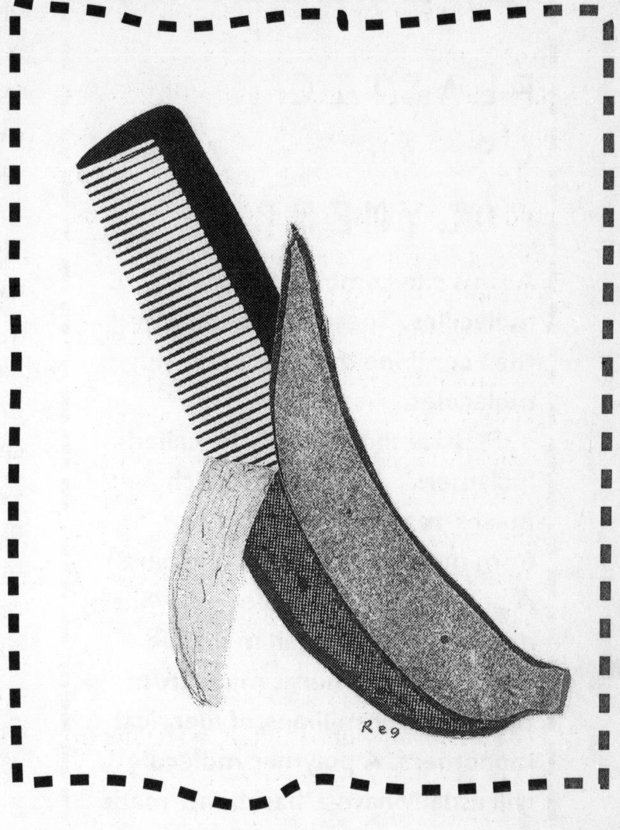

The great thing about plastics is that they're cheap to make from fossil fuels like crude oil, and that they last for ever. Unfortunately, these are also their great disadvantages. One day, the supply of fossil fuels will be so limited that plastics will be very expensive. And who wants a plastic cup that will outlast the pyramids? But recently, scientists have managed to persuade plants to make plastics for us — and biodegradable plastics at that.

From ivory to celluloid

Today, the plastics industry makes over 100 million tonnes of the stuff each year. In one scene from the 1967 movie *The Graduate*, Dustin Hoffman gets some unwanted advice from a concerned family friend. The well-meaning friend says: 'I just want to say one word to you. Just one word. Plastics. There's a great future in plastics.'

The plastics industry began to take off way back in 1868, after a rather rocky start around 1856.

In the middle 1800s, billiard balls were made from elephant ivory. Billiards became more popular, but elephants became more rare. Quite suddenly, there was a shortage

PLASTIC OR POLYMER?

Atoms can combine to make molecules. These molecules can then combine to make really huge molecules.

Plastics should really be called 'polymers'. A polymer (which means 'many parts') is made from many monomers ('one part'). A polymer is a big molecule, while a monomer is a small molecule. Often, the polymer is made from thousands, or millions, of identical monomers. A polymer molecule will usually have a 'backbone' made of carbon.

Ethylene (C_2H_4) is a gas that is lighter than air. But if you heat it to 200 Celsius degrees, and squash it with a pressure of 3,000 atmospheres, then it will turn into plastic with a very waxy feel. This plastic, or polymer, is called polyethylene.

Typical polymers are polyvinyl chloride (PVC), polystyrene, and polymethylmethacrylate (an acrylic, also called Perspex).

Some plastics are made up of repeating patterns of two (or more) identical sub-units. One such plastic is ABS, which is made from three monomers: acrylonitrile, butadiene and styrene. So ABS is called a 'copolymer'.

of ivory. In 1868, Phelan and Collander, a New England firm that made billiard and pool balls, offered the substantial prize of $10,000 for a cheap, and good, replacement for ivory.

An American inventor, John Wesley Hyatt, decided to enter this competition. He submitted a substance he called Celluloid.

But Wyatt had not invented celluloid. It had actually been invented 12 years earlier, in England, by Alexander Parkes, who was a chemist and inventor. Around 1856, he had combined nitrocellulose and camphor to make a flexible, hard and transparent material he called Parkesine. (Nitrocellulose is just cellulose that has been soaked in nitric acid — and yes, it is the stuff that goes bang!) Although Parkes got financial backing to market this material, nobody could see a use for it, and so it didn't sell. In 1868, he sold the patent rights to his 'useless' invention to John Hyatt.

Because the material was based on cellulose, John Hyatt called it Celluloid, and entered it in the competition. (We don't know whether he won the competition — some books say that he did, and others say that he didn't!) He registered the name, Celluloid, in 1872.

Celluloid soon became very popular. It was used to make billiard balls, knife handles, dental plates, dominoes, shirt collars and cuffs, combs, piano keys, and photographic film. In fact, the film industry could not have started without celluloid. Even today there is one use for celluloid for which we have not found a better material — table-tennis balls!

Celluloid is not completely artificial, because it is made from cellulose, which we get from plants.

Bakelite

The first completely synthetic plastic was made in 1907 by Leo Baekeland. Baekeland was a Belgian-born industrial chemist, who was lucky enough to get a very good education in the brand-new field of industrial chemistry, at the University of Ghent. He graduated, with great honours, at the tender age of 21, and taught at the university until 1889. Baekeland then went to the USA, and briefly joined a photographic company before setting up his own company. He invented Velox, a photographic paper which could be processed under artificial light. Velox rapidly became a fabulous success.

In 1899, Baekeland sold his company, and the rights to Velox, to George Eastman (the founder of Kodak) for a sum between $750,000 and $1,000,000. Back in 1899, that was a huge amount of money.

In 1905, he began looking for a substitute for the natural material called shellac. While experimenting in his laboratory, he mixed phenol (carbolic acid) and formaldehyde, and then heated it, to make a plastic which he named after himself — Bakelite. His notebook says: 'solidified matter yellowish and hard … looks promising … it will be worthwhile to determine how far this mass is able to make moulded materials … make a substitute for celluloid and hard rubber …'

By 1909, Bakelite was in full commercial production. It appeared everywhere: electrical plugs and sockets, radios, kitchen utensils, ice buckets, salt-and-pepper shakers, serviette rings and even jewellery. (Baekeland himself was so popular that he appeared on the cover of *Time* magazine in September 1924.)

HOW TO MAKE PLASTICS

Nature makes its own polymers, such as shellac and rubber.

We humans use two main methods to make a polymer: *condensation* polymerisation, and *addition* polymerisation. A third method has been recently invented: group transfer polymerisation.

In *condensation* polymerisation, you 'condense' some starting chemicals to make a final product. There are usually a few small molecules left over, such as hydrogen, water, glycol, or nitrogen. Bakelite, terylene, polyurethanes, polyesters and nylon are made this way.

In *addition* polymerisation, you 'add' some chemicals together to make a final product. All the starting chemicals are totally used up, and there are no chemicals left over. Polyurethane, polypropylene, polyvinyl chloride, polystyrene and polyethylene are made by this method. Addition polymers tend to be tens, or hundreds, of times bigger than condensation polymers.

Macromolecules

The next big jump in the plastics industry came from a German chemist. In May 1922, Hermann Staudinger published a scientific paper in which he proved that rubber was made from a long chain of identical isoprene molecules. In this paper, he became the first person to use the word 'macromolecule', meaning 'big molecule'. At the time, his paper attracted a lot of opposition and criticism. But he was right in his beliefs.

His idea, that little molecules can combine to make big molecules, was revolutionary. It meant that once you understood the basic chemistry, then it should be easy to make plastics with the properties that you wanted. In the past, chemists had just randomly mixed chemicals together to see what happened. Now chemists were working scientifically. Staudinger's paper led to a wave of scientific investigation that soon paid off in major breakthroughs.

As a result, during the 1920s and 1930s, new plastics came onto the market. They included polyvinyl chloride (PVC), used for electrical wire insulation, pipes, and siding sheets on houses; acrylic resin, used as an adhesive in laminated glass; urea-formaldehyde resins, used in electrical applications and tableware); and an improved version of celluloid.

Celluloid had a few disadvantages. It would gradually deteriorate in light, and it had another, far more dangerous, disadvantage — it was very inflammable.

But if you react celluloid with a chemical called acetic anhydride, you get cellulose acetate, which is not inflammable. Cellulose acetate was easy to shape into

THE BEAUTIES OF PLASTICS

Plastics have many good properties. They tend to be light, yet quite strong. They are often excellent insulators of both heat and electricity. They are also quite stable, and are resistant to corrosion by most acids, bases and solvents. They can often be degraded by UV light, which can be both an advantage and a disadvantage.

Unfortunately, apart from the reinforced plastics, plastics are not very stiff. They have a tendency to creep, or slowly distort, with time. And, apart from the formaldehyde plastics, they are not very hard. They tend to be brittle at low temperatures, and to lose strength and hardness at high temperatures.

What this means is that you have to choose the appropriate plastic for the conditions it will experience. And you can always add stuff — like dyes to block UV light, chemicals to change its properties, and fibres to make it stronger.

So for packaging, you might choose plastics such as polystyrene, polyvinyl chloride, polypropylene, and polyvinylidene chloride. Polyvinylidene chloride has great properties as a barrier to gases, so it stops the movement of oxygen. Polypropylene is not only good at stopping water vapour, it's also used as a fibre for ropes, and in carpeting.

thin, transparent sheets, which meant it could be used as a base for photographic film. Suddenly, photographic film was no longer a potential fire hazard, waiting for a tiny spark to set it burning fiercely. Cellulose acetate was also used in handles for toothbrushes and cutlery, telephone handsets and steering wheels.

Polymethylmethacrylate also appeared in the market. This transparent material is better known as acrylic, Lucite, Perspex or Plexiglas. Even today, it's used in glasses and cheaper camera lenses, neon signs, lighting fixtures, aircraft windows, and the tail-lights of automobiles.

Polystyrene resins were invented around 1937. They're very resistant to the effects of chemicals and low temperatures, and are very resistant to water. They found an immediate use in electronics (as radio-frequency insulation) in low temperature applications (refrigerators, high-altitude aeroplanes, etc.).

Teflon

Polytetrafluoroethylene, better known as Teflon, was accidentally invented on 6 April 1938. Roy J. Plunkett Jr. had been experimenting with a gas called tetrafluoroethylene. He was interested in finding a working gas to use in refrigerators. (The working gas is the stuff that goes around and around, changing from liquid to gas and back again. Whenever it changes from a liquid to a gas, it 'sucks' heat from the surroundings.) He opened the valve on an unused cylinder of tetrafluoroethylene, but nothing came out. Most people would simply have got another cylinder of this gas from the company store, but Roy had a mind brimming with curiosity.

First, he weighed the cylinder, and found that it had the weight of a full cylinder of gas. So it had something inside. Then, he ran a thin wire through the valve, which meant that it was open. Finally, he hacksawed the tank open! Inside he found a white, slippery, waxy powder. Somehow, the gas, tetrafluoroethylene, had spontaneously reacted with itself to turn itself into a solid, polytetrafluoroethylene. He soon found how to make this wonderful powder.

And it *was* wonderful. Teflon was the most inert plastic he had ever seen — it was unaffected by heat, it would not react with anything (acids and bases would not touch it), nothing could dissolve it, and it was incredibly slippery (more slippery than wet ice on wet ice). But it was expensive, and so for a little while, it had no commercial use.

A few years later, the top-secret Manhattan Project was busy trying to make the first atom bombs. Part of this process involved turning uranium into a gas, uranium hexafluoride. Unfortunately, uranium hexafluoride would corrode almost anything that it came into contact with, including metal parts in pipes. This was a major problem. But cost was not a problem in the Manhattan Project, and once the military heard about Teflon, they bought huge quantities of it.

Because Teflon is very biocompatible, the human immune system ignores it. So it's used in artificial heart valves and artificial aortas, as well as artificial corneas, and artificial bones (ear bones, chin, fingers, nose, skull, and various joints). It's tough enough to withstand the harsh conditions of outer space, so it's used in space suits.

EXPLODING PERSPEX CURED CATARACTS

During World War II, polymethylmethacrylate (acrylic, Lucite, Perspex or Plexiglas) was used to make windscreens in planes. It was light, strong, and didn't shatter as easily as glass. However, it would shatter if it was hit by bullets. As a result, many pilots and gunners on planes ended up with variously sized pieces of acrylic inside their eyes.

To the surprise of the eye doctors, the injured eyes did not reject this foreign material. The acrylic just sat there in the eye, without any reaction, and the immune system ignored it. This was how modern medicine found out that acrylic could be used as an implant material in the human eye!

The human eye is roughly the size of a golf ball. The incoming light is bent twice, to bring it to a focus on the retina at the back of the eye. Most of the bending is done by the cornea at the front of the eye. Further bending of the light is done by a part of the eye called the lens. The lens of the eye actually looks like the lens in a simple magnifying glass. As people get older, they are more likely to have their lens gradually turn opaque. This disease is called 'cataract', and is the most common cause of blindness in the world.

An eye doctor can remove the slightly opaque lens, and replace it with a lens made of acrylic. Because this artificial lens sits inside the eye, it's called an 'intra-ocular lens'.

The great Professor Fred Hollows's Magnificent Dream was to make these lenses available to the poor countries of the world. His idea was that the lenses were to be made in these poor countries, and then installed by local doctors of these countries. This humanitarian project is currently being carried out by the Fred Hollows Foundation.

And it all was made possible because somebody in World War II noticed that the human eye did not reject acrylic.

Nylon

Nylon (polyamide), the first high-performance engineering plastic, was also introduced during the 1930s. Nylons are strong, resistant to abrasion, easy to machine, and have low friction. This is why they're made into lubrication-free, quiet, machine parts such as small gears, bearings, slides, and rollers. Nylon became famous in World War II because of its use in parachutes and stockings.

(By the way, 'denier', the weight that is used to measure the fineness of stockings, is the weight, in grams, of 9 kilometres of thread. So 20 denier stockings are made from a fibre that weighs 20 grams per 9 kilometres.)

During World War II, many sources of natural materials became unavailable, so the plastics industry had to come up with substitutes. They were able to make everything from textile fibres to synthetic rubber.

After World War II

After World War II, there was another boom in the plastics industry.

In some areas, plastics were able to compete successfully with metals. Polycarbonate became available in 1956, and was used to make bulletproof 'glass'. Polycarbonate is also resistant to heat and flames, and is used in power-tool casings, hard hats, and vandal-proof glazing.

Two of the most important plastics developed were polyethylene and polypropylene. Polyethylene was developed in 1953 by the German chemist Karl Ziegler, and polypropylene was developed in 1954 by the Italian chemist Giulio Natta. In 1963, Ziegler and Natta shared the Nobel Prize in Chemistry for their great achievements in plastics.

What is plastic?

The word plastic comes from the Greek word *plastikos*, which means 'to mould' or 'able to be moulded'. So a plastic is a substance that can be moulded, by heat and pressure, into the shape you want.

Plastics are made of long chains of chemicals called polymers. These long polymers can be slippery and skinny (like spaghetti), or clingy and heavily cross-linked (like barbed wire).

There are two types of plastics: thermoplastics and thermosets. These are fairly 'soft' categories, with a lot of overlap. Thermoplastics can be made to have a wider range of useful properties than thermosets, so industry makes more thermoplastics than thermosets.

Thermoplastics

A thermoplastic softens when it heats up, and hardens when it cools down – and it can go through this cycle of hot–soft/cold–hard repeatedly. Thermoplastics are made from long slippery chains, which are tangled up together at room temperatures. But at high temperatures, these chains slip loose from each other, so making the thermoplastic soft. As the thermoplastic cools down, the polymers get tangled again, and the plastic becomes hard. Typical thermoplastics are acrylic (also called Perspex), ABS (used in suitcases), polyvinyl chloride (PVC, used in shower curtains, electrical insulation and garden hoses) and polyethylene.

By weight, the most popular plastic made worldwide is polyethylene. This thermoplastic is made in five grades: low-density (LDPE), medium-density (MDPE), high-density (HDPE), ultrahigh molecular weight (UHMW), and irradiated (polyethylene that now has internal molecular cross-links, thanks to radiation). Polyethylene is even made as a flexible foam.

Thermoplastics are used to package many foods. By 1993, over 5 million tonnes of low-density polyethylene (LDPE) were used to package food each year in the USA. You'll mostly see LDPE as rolls of clear plastic cling-wrap. High-density polyethylene (HPDE) is used to make garbage bags, containers, and household plumbing pipes.

PVC, another thermoplastic, is the second most popular plastic. It's used in window frames, floor tiles, and even household guttering.

Most thermoplastics, apart from the acrylic polymers and the fluorinated polymers, are degraded by ultraviolet (UV) radiation. If the final colour of the product doesn't matter, carbon black is usually added to absorb the UV. If the final product needs to be a colour other than black, other more expensive additives are used.

Teflon is a thermoplastic that cannot be softened by heat. This is because it has a rather complicated molecular structure.

Thermosets

The other type of plastic is called a thermoset. It has strong cross-links (like barbed wire) between the polymer chains, so once it sets, it remains hard. Actually, some thermosets that are lightly cross-linked will behave like thermoplastics — they will soften when heated, and harden when cooled. But in general, most thermosets will harden when they are heated. This heating sets off a final and permanent cross-linking, which 'fixes' a 'true' thermoset.

Typical thermosets are Formica, Melamine, alkyd resin (used in screwdriver handles), Bakelite, and epoxies. For some reason, wood, wool and paints are also included as thermosets. The cross-linking of thermosets tends to make them more rigid and brittle than thermoplastics.

Plants versus petrochemicals

The very first plastic, celluloid, was made from a plant product – the cellulose came from cotton. Even today, one type of nylon (Nylon 11) is based on the oil from castor beans. But most plastics come from petrochemicals. Today, petrochemicals are easy to get, and cheap. But this will very probably change over the next century.

Most plastics are not environmentally degradable. A few biodegradable plastics have been developed, but most don't degrade in the low-oxygen environment of a landfill. This is part of the interest in recycling. Plastics such as polyethylene terephlalate (PET), which are used in bottles holding sweet fizzy drinks, are easy to recycle.

So you can see that there are advantages in being able to make a biodegradable plastic from something other than petrochemicals.

This was part of the thinking behind the research to modify a plant to make a thermoplastic. It was carried out by Chris Somerville and his colleagues (then at the

NATURAL POLYESTER

Agracetus Inc. — a biotechnology company in Middleton, Wisconsin — has managed to genetically engineer a cotton plant to make PHB in the centre of its cotton fibres. The PHB appears as granules, not as fibres.

The company is trying to make a wrinkle-free or wrinkle-resistant cotton, by natural means. Maybe these cotton fibres with a plastic core will give them the answer.

Department of Energy-Plant Research Laboratory at Michigan State University), and Douglas Dennis (Department of Biology at James Madison University in Harrisonburg, Virginia).

TEFLON AND THE SPACE RACE

In the 1960s, the USA and the USSR were engaged in a Space Race to the Moon. Teflon entered the commercial market in the 1960s, when it was used to coat frying pans. Perhaps the timing of these two events led to the claim: 'The only good thing to come out of the Space Race to the Moon was Teflon on frying pans.'

This claim is completely false. First, Teflon was invented in 1938 and was in use 20 years before the 1960s. Second, every $1 that was invested in the American Space Program has paid back at least $3 in commercial dividends. Thousands of good things have come out of the Space Race.

Besides, during the years of the space program, the amount of money spent on the program was less than the amount of money spent by the American public on running-shoes!

They were particularly interested in a thermoplastic called polyhydroxybutyrate (PHB). Some bacteria use PHB as a storage material. In the same way that we humans store extra food as fat, so a few bacteria store their extra energy as PHB plastic. It's an energy source for them to call upon when times are lean. These bacteria are able to easily break down the PHB. In other words, PHB — which the industrial chemists call 'a high molecular weight polyester thermoplastic' — is actually biodegradable by bacteria.

It turns out that both fungi and bacteria can break down PHB into water and carbon dioxide. If PHB is exposed to air, half of it will have decayed within six weeks – an astonishingly fast time.

ICI, the biggest chemicals company in Britain, has been harvesting bacteria since 1991 to make PHB. The company transforms the PHB (which ICI calls 'Biopol') into bottles. These bottles were at first available only in Germany, where the Wella hair-care company used them to package shampoo. The production cost of PHB was around $30 a kilogram from crude oil, and a little bit cheaper from bacteria.

Chris Somerville and his colleagues believed that if they could find a way to get plants to make PHB, they could bring the cost down to $1 or $2 a kilogram. Plants are really good at making large quantities of stuff, and they're cheap to run – living on sunlight, water and air.

One bacteria that makes PHB is called *Alcaligenes eutrophus*. It starts off with a common biological chemical called acetyl-coenzyme A. The bacteria then performs three chemical reactions on the acetyl-coenzyme A, until it gets turned into PHB.

Chris Somerville and his colleagues knew that a plant called *Arabidopsis thaliana*, or thale cress, was able to do one of these three reactions. (Thale cress is a member of the mustard family, and is related to cauliflower and broccoli.) So they did a bit of clever genetic engineering. They found exactly which genes inside the bacteria were the ones that actually controlled the other two reactions. They 'snipped' out these genes, and then implanted them into a virus that infects cauliflowers. Their tame and modified virus then infected the thale cress, and successfully inserted the genes into it.

They succeeded in their ground-breaking work in 1992. In the bacteria, the PHB formed tiny balls about 0.2 to 0.5 microns (millionths of a metre) across. In the plant, they found similarly sized granules of PHB. This attracted a lot of media attention. Chris Somerville said: 'For the first time, a plant has been genetically engineered to make plastic.' But in those early days, they had a few problems.

First, the production rate was very low. They managed to get a concentration of PHB of only 0.01 per cent in the living plant. Second, the plants became sickly, and grew to only 20–45 per cent of their normal weight. This was probably because the plants were diverting carbon away from making plant tissue to making PHB. Third, the amount of seed that the plants made dropped to about 35 per cent of normal.

But Chris Somerville and his colleagues persisted. He said, in an interview with Andy Coglan in the *New Scientist*: 'Now, we want to find how much we can improve it. We need to raise production 100-fold, but I think that's feasible.'

Chris Somerville is now a professor at the Carnegie Institute in California. He has spent the years since his initial announcement doing some very careful breeding. In September 1995, his team announced that they have managed to increase the yield of PHB to 14 per cent of the dry weight of the plant.

So, one day soon, your plastic indoor plants could very well be made out of real plants!

REFERENCES

'Biodegradable plastic hits the production line', *New Scientist*, no. 1715, 5 May 1990, p. 18.

Coghlan, Andy, 'How to sow cress and reap plastic', *New Scientist*, no. 1820, 9 May 1992, p. 20.

Encyclopaedia Britannica (Electronic Version), 1996.

'Introducing the plastic that grows in weeds', *Focus*, September 1995, p. 20.

'100% natural polyester', *Science*, vol. 266, 16 December 1966, p. 1811.

'Plant factory', *Science*, vol. 256, 24 April 1992, p. 419.

Poirier, Yves, *et al.*, 'Polyhydroxybutyrate, a biodegradable thermoplastic, produced in transgenic plants', *Science*, vol. 256, 24 April 1992, pp. 520–523.

ANCIENT INFORMATION HIGHWAY & FAX

We humans all want to communicate, and the usual way is to talk with each other. But today we have the Internet, which is also laughingly called the 'Information Super-Highway' — laughingly because it's 90 per cent mis-information, and because it's not even a highway, much less a super-highway. The Internet (also called 'the net') carries enormous amounts of information, which travel on beams of light inside optic fibres. It might seem that this technology is brand new. But the first information highway started up about two centuries ago, and it also used light! And even today, the fax machine still runs on light.

SMELLY MACHINES

Back in the days of the Group 1 and Group 2 standards, the fax machines used burning chemicals to make an image. These chemicals made quite a smell, so unscrupulous sales people would actually smoke a cigar while demonstrating their machine, to hide the smell of the machine!

The phone, the fax and the net are just symbols of the universal desire to communicate.

Honeybees let each other know where the nectar is by waggling their bodies in a rhythmic pattern and pointing to the Sun. Many insects communicate by special smells or odours, called pheromones. Animals attract, or repel, each other with gestures and smells. Women use smells to synchronise their periods. Our ancient ancestors made rock carvings and cave paintings to pass on messages.

Communication has always been absolutely essential to the military. Tom-tom drums were used in Africa, and carrier pigeons have a long history. The ancient Persians and Romans used riders, who picked up a fresh horse at each station, to relay messages. In fact, the Roman word for 'station' is *positus*, which gives us the modern phrase, 'postal service'.

Three thousand years ago, the Greeks who were attacking Troy lit fires on hilltops to let the people back home, in Argos, Greece, know that Troy had finally fallen. This news, via a series of bonfires, took just a few hours to cover some 800 kilometres. And as recently as 1588, the English used the 'fires-on-hilltops' method to send a message from Plymouth to London, some 300 kilometres, in just 20 minutes, to let Londoners know that the Spanish Armada was on its way.

These bonfires used light to send information. But the trouble with the 'fires-on-hilltops' method is that it can send only a few units of information — and usually just in one direction.

Today's information highway

Today's information highway runs on light. Many of us use the net to email each other, to communicate in groups with each other, and to search for information. The core of the net sends information via optic fibres, but most of the peripheral links to the net are old-fashioned copper wires (such as the telephone wires to your modem).

The information super-highway is a highway *only* if you have direct access to the high-speed optic fibre — but this usually happens only in a university or a hospital. Working with optic fibre, you can sometimes download information at a million bits per second. But if you have to download information via ordinary copper wires, you're sucking through a very narrow straw indeed. You usually download information about 100 times slower — about 10,000 bits per second.

For the average user, working from home via noisy copper telephone lines and a modem, the net is not a super-highway, but a muddy dirt track. Average users don't cruise in sports cars, but bounce along on pogo sticks — which sometimes get stuck in the mud (when the telephone line suddenly 'drops out'). And finally, the road signs are not in English, but in some obscure mixture of Unix, Latin and Martian!

Even so, the amount of traffic on the net is growing rapidly. One part of the Internet is called the World Wide Web. It took the whole of 1992 for the Web to carry 500 megabytes of data — roughly equivalent to a 30-volume encyclopaedia. In 1993, it took nine days to carry that much data, and in 1994, it took 18 minutes!

The ancient Greeks, and more recently the English, used the light from fires to carry their messages. The 'fires-on-hilltops' method was slow, it didn't carry much information, and it carried that information in only one direction. Even so, it was a genuine optical information communication link.

In 1684, Robert Hooke put forward his ideas about an optical telegraph. They were published in 1726. But it was a French physicist and engineer, in the 1790s, who actually built the first optical data network.

JAPANESE LOVE FAX

The international telegraph code allows the transmission of only some 56 characters. This was just fine if you used the English alphabet — 26 letters, plus 10 numbers, plus odd characters like $, %, + and so on.

But the Japanese language does not use a small number of letters — it uses a large number of pictographs, or characters. In fact, the average, well-educated Japanese person would know at least 2,000 characters. A telegraph system with just 56 possible characters was simply not suitable. That's why the Japanese were keen to develop the fax machine, with its ability to send pictures. Even today, the Japanese make practically all the fax machines in the world.

18th-century information highway

In the late 18th century, scientists didn't understand electricity well enough to use it to send information, but they did understand telescopes and signal flags. At 11:00 am, on 2 March 1791, Claude Chappe used his first optical telegraph to send a message some 16 kilometres. The hardware was the combination of a telescope and a large clockface. The hand on the clockface could be pointed at various angles. The software was a special code-book where various angles of the hand meant various letters, words and even phrases. The first message, which took four minutes to send, was: 'If you succeed, you will soon bask in glory.'

But nothing much happened until 1793, by which time, the new Republic of France was well and truly at war with most of its neighbours. France was surrounded by unfriendly neighbours — Britain, Holland, Prussia and Spain. Even worse, France faced internal revolt from the towns of Lyons and Marseilles. Finally, the British fleet had occupied Toulon. The situation

A
D
C
B
a
b
c
d
©P.POUND 96

looked hopeless. But France had one thing in her favour — her enemies did not have a good communication system set up between themselves.

France did, however, possess fast communications, and the military advantages were obvious. A new word was invented to describe it: *telegraph*, which means 'far writer'. Claude Chappe redesigned his first optical telegraph. The second design did not use a clock. It used a long wooden beam (which could be tilted to various angles), with a short arm at each end of the long arm. It looked a little like a human with out-stretched arms (the long beam), with a signal flag (the short arm) in each hand. Information was coded by the angles of the beam, and the angles of the arms. These optical telegraphs were mounted on towers, up to 32 kilometres apart.

The first long line ran some 190 kilometres, from Paris to the border city of Lille. This line used 15 stations. On 15 August 1794, the first official message on this line announced good fortune for the French — their troops had retaken Le Quesnoy from the Austrians and the Prussians. On 30 August, this line again carried good news — the recapture of Condé. Chappe's optical telegraph looked better as it continued to carry more good news, as French troops successfully advanced north into Holland. The French government decided, on 3 October, to fund another optical telegraph line from Paris to Landau.

When Napoleon Bonaparte seized power in 1799, there were already some 150 relay stations covering hundreds of kilometres, but he ordered further expansion of the network. He realised the military significance of good communications. By 1804, a line ran some 720 kilometres from Paris in France, to Milan in Italy. A year later, there were lines running north, south, east and west from Paris. Napoleon was so impressed with the military advantages of this early information network that he took mobile stations with him on his unsuccessful invasion of Russia in 1812.

Those optical semaphore telegraph stations were quite speedy. The relay stations would hold each position of the beam and arms for 20 to 30 seconds to make the signal easy to recognise. The average length of an entry in the code-book was about 10 letters. So, way back in 1800, the effective data transfer rate was about 20 letters per minute. It was only in 1837 that the newly invented 'electric telegraph' could equal or better that transfer rate. Part of the reason for the slowness of the transmission was that each station would repeat each signal, to be sure that the signal was correct.

In 1852, the optical telegraph had 556 relay stations on some 4,800 kilometres of lines in France, and linked 29 French cities to Paris. But the writing was on the wall for the old-fashioned optical telegraph, and by 1858, the electric telegraph was sending information at 2,000 letters per minute. Even so, the last optical telegraph stations in Europe closed as late as 1881.

Data compression and error correction

The optical telegraph was not a dead-end. Two centuries ago, the builders of those early optical information highways invented data compression and error correction, and they were using sophisticated methods of encryption.

The first code-book in 1791 had 9,999 entries. The commonly used entries were given low numbers. This simple technique was an early ancestor of modern data-compression techniques. By 1799, the code-book had over 25,000 entries. With the increase in entries came increasingly sophisticated data compression.

In the software, those early information optical network designers came up with secret codes to encrypt the data, special techniques to average out errors, and sophisticated methods of compressing the data.

Fax machine

Isn't it strange that, two centuries later, we have once again seen the light and gone back to optical methods to talk to each other?

One of these optical methods is the fax machine. 'Fax' is short for 'facsimile transmission'. In the last 10 years, fax machines have become incredibly popular. But they were actually invented about 150 years ago — well before the invention of the telephone!

Ancient fax technology

Many people worked on the concept of the fax machine. In 1826, Sir Humphrey Davy proposed the idea of a fax machine. He even managed to send electrical signals a short distance, but he did not come close to actually building a fax machine incorporating a transmitter and receiver.

One of the earliest to do so was Alexander Bain, a Scotsman, who invented the first electric clock. He was born in 1810, in the days when hardware stores were quite rare. So experimenters had to make practically everything themselves. The story goes that not only did he make his own batteries by burying metal plates in the ground, but he also used sprigs of heather as primitive springs — and for hinges he used the jaw bones of cattle! He actually sent the first fax in 1842, but patented his facsimile machine on 27 May 1843 (Patent No. 9745). This was 33 years before Alexander Graham Bell patented the telephone.

For his fax machine, Bain used the same principle that is used today both in modern fax machines and in TVs.

If you stand a few metres away from a TV, the picture looks smooth. But if you get up close, you can see the individual dots that make up the picture. Each dot is called a 'pixel' (short for 'picture element') or 'pel'. On a TV screen, there are about 1,000 dots horizontally (across the screen), and about 600 dots vertically (up and down the screen).

They are written onto the TV screen by an electron gun, which writes the first line by starting at the top left and scanning across to the top right. It then jumps down by a distance of one dot, and back to the extreme left, and then paints another 1,000 dots onto the screen. The gun then continues this process until it gets to the bottom of the TV screen, having written some 600 lines in 1/25th of a second. Then the electron gun writes another 600 lines in the next 25th of a second.

Bain used the same left-to-right and top-to-bottom scanning method in his fax machine.

OPTIC FIBRE

Optic fibres are just long strands of very pure and very transparent coated glass. Ordinary window glass seems quite transparent. But if you have ever seen several panes of window glass stacked together, you might have noticed that the glass has a strong greenish tinge — with less than 30 centimetres total thickness of glass. The glass from which optic fibres are made is transparent, not for centimetres, but for 20 kilometres! The glass in an optic fibre is very skinny — between 0.1 and 0.2 millimetres in diameter.

There are two main types of optic fibre: mono-mode and multi-mode. The mono-mode fibres carry only a single signal. But they can carry this signal 120 kilometres before a repeater amplifier is needed to 'pump up' the signal. The multi-mode fibres can carry many signals at the same time, but need repeater amplifiers after just 10 kilometres.

Optic fibres can carry much more information than copper wires, because they carry beams of light. The frequency of light is around 100 trillion (i.e., 100 million million) Hertz (cycles per second). A recognisable voice communication covers about 5,000 Hertz. If you allow a 'buffer' of 2,500 Hertz on each side, a voice signal takes up 10,000 Hertz. Theoretically, you can squash 10 billion (i.e., 10 thousand million) voice channels into a beam of light. Practically speaking, 40,000 voice channels fit easily.

The first transatlantic optic fibre cable, TAT–8, was laid in 1988. It carried 40,000 conversations — which was three times more than could be carried by all seven of the undersea copper cables that existed at the time!

But optic fibres have other advantages besides their enormous carrying capacity.

They are much lighter than copper wire, and very skinny. This is important if you are trying to wrestle with a big drum loaded with a few thousand kilometres of cable. These properties are also very useful in planes.

Optic fibres are also immune to electrical interference. This makes them useful in electrically noisy environments, such as electrical power plants, factories and hospitals.

At the transmitter, he arranged standard lead printers' type to make up a message of a few words.

When printers prepare metal type for printing, using old-fashioned pre-computer technology, they usually place a thin layer of ink on the type. The ink covers the raised sections of the type, but there's not enough of it to run into the hollows. When the type is pressed against paper, a clear imprint of the letter is transferred to the paper.

But Bain did not use ink. Instead, he used electricity at the transmitter to read the page, and chemistry at the receiver to write the page.

At the transmitter, he slowly scanned a metal point across a row of printers' type

OTHER OLD OPTICAL NETWORKS

Many countries used optical networks. There were quite a few variations. Some used signal arms that could be tilted to different angles, others used flat boards that could be tilted up on edge and so become effectively invisible, while others even used mirrors to reflect the sun.

By 1794, news of Chappe's optical telegraph reached England. Lord George Murray proposed a system that used a large wooden board with six holes. One or more of these holes could be closed to give different messages. Beginning in 1796, the Admiralty erected a chain of these towers from London to Deal, Portsmouth, Yarmouth and Plymouth.

This optical telegraph could work very rapidly indeed: 'A single signal has been transmitted to Plymouth and back (London) in three minutes, which by the telegraph route is at least 500 miles. In this instance, however, notice had been given to make ready, and every captain was at his post to receive and return the signals. The progress was at the rate of 170 miles in a minute, or 3 miles per second, or 3 seconds at each station: a rapidity truly wonderful!'

However, the line was expensive in terms of people to run it, and could be shut down by darkness, heavy rain, and even smog! 'The Station on Putney Heath, communicating with Chelsea, is generally rendered useless during easterly winds by the smoke of London which fills the valley of the Thames.' The Plymouth line was closed down in 1847.

Germany and Russia also had optical telegraphs.

In 1800, Jonathan Grout built the first optical telegraph in the USA, joining Boston to Martha's Vineyard. The main purpose of this 104-kilometre-long line was to send information about shipping.

In Sweden, Abraham Edelcrantz began construction of an optical telegraph on 30 January 1795. By 1809, the Swedish network used some 50 stations to cover a distance of about 200 kilometres.

letters. The metal point was part of an electrical circuit. When the point touched the raised part of the lead letter, it completed the electrical circuit. This created an electrical signal, which was sent off to the receiver. When the point was over a 'valley' in the lead type, it did not complete the circuit, and no signal was sent to the receiver. When the point got to the end of a 'line', it zipped back to the beginning of the next line. There were several scanning lines in each letter.

Wires carried the electrical signal from the transmitter to the receiver.

At the receiver, a chemical process was happening. A metal point was moving across a sheet of paper soaked in potassium ferrocyanide. Whenever the metal point at the distant transmitter made electrical contact with a letter, at the receiver electricity jumped from the metal point onto the chemical-soaked paper, and turned it black!

The metal point at the receiver had to scan in exact synchrony with the metal point at the transmitter. It used watch-maker technology to control how long it took for the left-to-right scan, and how big each downward step was.

First, both metal points (transmitter and receiver) had to scan across at the same rate. Second, both metal points had to jump down to the next 'line' at the same time, and had to jump down by the same amount. This was made possible by Bain's skill as a clockmaker. Modern tests show that if the scanning rates differ by as little as one part in 10,000, you immediately get gibberish at the receiver. Bain was a very careful, and precise, worker.

By 1846, he was able to send information at 253 words per minute — which was the world speed record for several years. He eventually went bankrupt after a series of very nasty patent battles with Charles Wheatstone in England and Samuel Morse in America, who were both working on similar devices.

Bain's machine was splendidly ingenious, but it could not survive for financial reasons. This was because it sent pictures, not words. The telegraph using the Morse Code could send more words in each minute than the fax machine could. Of course, the fax machine could send pictures, which the telegraph could not. But at that time, nobody really had a need to send pictures instantaneously.

The first commercial fax service began around 1865, linking the French cities of Paris and Lyons. It still used mechanical watchmaker-technology to scan the page, three lines in each millimetre. It also used electricity to read the page, and chemistry to write the page. In its first year of operation, some 5,000 pages were sent. It used the 'Pantlégraphe', which was designed by an Italian professor of physics, Giovanni Caselli. (An earlier and more limited service, using Caselli's machine, had been installed in 1856, running between the French cities of Paris, Amiens and Marseilles.)

Modern fax technology

The major breakthrough in fax technology came in 1878. In that year, it was discovered that the electrical resistance of an element called selenium changed when light shone on it. So this method used light to read the page.

Consider a white page with black print on it. It's white where there is no print, and black where there is print. A beam of light, arranged as a very thin horizontal line, was scanned across the page. White paper would reflect a lot of light, while the black print would reflect hardly any light at all. This reflected light then fell on a selenium electric cell, which varied its resistance according to how much light fell on it. This varying resistance was turned into an electrical signal, which was sent to the receiving machine. This meant the end of the electrical reading systems, as the optical reading systems came into widespread use.

The electrical reading systems worked only with a message that had been converted to a metallic form, such as lead letters arranged in a row. The optical reading systems could read anything that had a difference between light and dark, such as the printed page or a photograph.

The first photographic fax machine was invented in 1906 by a German physicist, Arthur Korn. It incorporated three advances over the original Bain fax machine. First, it used the optical reading method to read a photograph. Second, instead of the clockwork-technology to scan the photo, it used synchronous electrical motors. Third, at the receiving station, it passed a beam of light over photographic paper, which was then chemically processed. In 1907, the first photograph was sent through electric wires between Munich and Berlin.

Around the mid–1920s the quality had improved so much that newspapers began to print faxed photos. But the standard telephone lines were much too noisy, so special lines had to be used. The machines were a few metres high, so that a transmitting machine and a receiving machine would completely fill a room!

By 1924, American Telephone & Telegraph (AT&T) was regularly sending photographs by wire from Chicago, and Cleveland, Ohio, to New York. These photos, with a resolution of 100 lines per inch (about four lines per millimetre) were good enough to publish in newspapers. But the process was slow — a photo 125 millimetres by 175 millimetres took seven minutes to send. Even so, it was a lot faster than the postal mail. These pictures were called 'wire photos'. This was the first common, and practical, use of facsimile technology.

During World War II, the German Army used the Feldfernschreiber Portable Field Telegraph (also called the Hellschreiber). This device combined a teleprinter with a fax, and so it could send and receive both text and pictures.

By 1948, Western Union had improved the technology with its 'desk-fax' service. By the time this service was discontinued in the 1960s, some 50,000 machines had been built. Other companies were also interested. In Germany in the 1950s, Magnavox developed their own fax machine, as did Xerox in the USA in 1964.

But even though different manufacturers had made various models of fax machines, they all shared one common, and major, problem — they were not compatible with each other. The fax machines of different manufacturers could not fax messages to each other.

In 1968, the Group 1 Fax Standard was introduced, which would send a full A4 page in about six minutes. The resolution was about 100 lines per inch, or four lines per millimetre. Its standards were not 'compulsory', but only 'recommended'.

FAX - BIG & LITTLE

According to the 1993 *Guinness Book of Records*, the biggest fax machine in the world is made by WideCom Group, of Ontario, in Canada. One of its machines, the 'WIDE-fax 36', can scan, transmit and print sheets up to 91.4 centimetres wide.

The smallest fax is made by Real Time Strategies Inc. Its Pagentry model is 75 by 125 by 18 millimetres in size, and weighs 142 grams.

So the situation of incompatibility was slightly improved — but there were still problems. It was possible to send faxes within France, but impossible to send faxes out of, or into, France! You could send a fax from Europe to the USA, but not from the USA to Europe!

In 1976, the new Group 2 Standard was introduced. This time around, everybody involved realised how important it was to have international standards. The transmission time for an A4 page was dropped down to three minutes, by increasing the speed of data transmission. It was now possible to send a fax anywhere in the world, fairly quickly. But the price was still high, and the quality was poor, so the consumer market was not interested in fax technology. It was used mainly by banks, the military, police departments, publishers and news agencies. In 1976, there were only some 69,000 fax machines in the USA.

The Group 3 Standard (the one which we all use today), was set up in 1981. By increasing the transmission rate to 9,600 bits per second, an A4 page could now be sent in 20 seconds. The running costs were lower, and gradually the purchase price came down. The consumers responded, and by 1982, there were 350,000 fax machines in the USA.

Group 3 fax machines scan across the page with 1,728 tiny light sensors. They scan downward by advancing the paper by one-quarter of a millimetre. (The numbers 1,728 and one-quarter of a millimetre were chosen to give a resolution close to 100 dots per inch.) If you choose high resolution, the step size drops by half, to one-eighth of a millimetre. So, in high-resolution mode, the paper takes twice as long to be scanned by the fax machine.

Most fax machines work with a special thermal paper, which turns black when it's heated. You can prove this easily for yourself. If you run your fingernail rapidly across thermal paper, the friction between your fingernail and the paper generates enough heat to leave a black mark on the paper. The thermal paper also turns black if you place it close to an electric heater.

Inside your fax machine, there are 1,728 tiny electric heating elements, which are arranged in a thin line. The special thermal paper runs very close to these heating elements. If they are switched on, they heat up, and blacken, the paper. Because these elements are so tiny, they can heat up and cool down very rapidly indeed — about 300 times per second. This is one of the limits to the speed at which a thermal-paper fax machine can operate.

The resolution of thermal paper can't be better than eight dots in each millimetre. This is because the heat spreads through the paper and makes the dots merge into each other, so reducing the sharpness. Thermal paper has a few other disadvantages — it tends to curl up, and if you use a highlighter pen on thermal paper the text will vanish. In fact, if you just store the fax message in your filing cabinet, the text will fade, and vanish, after a few years. But now there are plain paper fax machines, which work either with a laser printer, or with bubble-jet devices. These are more expensive, but they have a sharper image (up around 14 dots per millimetre, or 360 dots per inch).

When another fax machine rings your fax machine, they go through a process called 'The Handshake'. This happens at the rather slow speed of about 300 bits per second. During the handshake, not only do

DEATH OF THE FAX

Many fax machines 'die' before they wear out — and this death is often caused by a lightning bolt.

A fax machine is connected to two external lines — the electric power supply line, and the telephone. This means it has more chances of getting 'zapped' if there's a lightning storm in your area. When a lightning bolt hits a line, it sends an electrical 'spike', some 5,000 volts high, down the line. The circuits in your fax machine, which run on less than 10 volts, simply fry and burn out. This spike is very tall, and lasts for only a few millionths of a second.

For some unknown reason, devices that block this spike are called 'surge protectors' — even though a surge is thought of as a slow and small increase in voltage (say from 240 up to 270 volts, over a period of a few seconds or minutes). But they have been sold as surge protectors for so long that it's too late to change the name to 'spike protectors'.

When I bought my second-hand fax machine, I sent it to the manufacturer's service department to be checked out. The service people promised that it should be ready in a few days, and that they would ring me. When they didn't ring after a week, I rang them to enquire as to when it would be ready. They replied that the service department was full of dead fax machines. The technician said: 'You know that big electrical storm that hit Bankstown a few days ago? Well, you can plot the path of the storm by the addresses from which all these dead fax machines came!'

So when my fax machine came back, I bough surge protectors (which should really be called spike protectors). I installed two of them on the power line (one at the fuse box, and the other near the fax machine), and another one on the telephone line.

they let each other know what features, options and extras they each have, but they also test the line to work out the highest speed they can send data at. If the line is good enough to transmit at 9,600 bits per second, the fax machines will operate at that speed. Otherwise, they will drop down to 7,200, 4,800, and even 2,400 bits per second.

Once the handshaking is over, the sender begins to transmit the message. In Australia, because our phone lines are so good, about 90 per cent of all fax messages travel at the top Group 3 rate, which is 9,600 bits per second. But if the line becomes suddenly noisy half way through the transmission, the fax will slow down so that the signal can still get through.

These days, fax machines are being used for everything. Many people have formal correspondences on them, and use them instead of the post (now called 'snail mail'). Some people use fax machines to send their lunch orders to their local corner shop. Many farmers use their fax machines to download weather maps from the weather bureau. When Boris Yeltsin was confined to the Russian Parliament, he kept in contact with the outside world via fax machines. And in 1989 the Chinese students of Tiananmen Square also used fax machines to let the outside world know what was really happening.

Now, there's one very important thing about all this new technology: it was supposed to give us the paperless office. But fax machines use and generate enormous amounts of paper, and the paperless office is beginning to look as far away as the paperless toilet.

REFERENCES

Holzmann, Gerard J. & Pehrson, Björn, 'The first data networks', *Scientific American*, January 1994, pp. 112–117.

How Is It Done?, Reader's Digest (Australia), 1990, pp. 77–78, 141, 212–215, 224–225, 228, 230.

Hunkin, Tim, 'Just give me the fax', *New Scientist*, no. 1860, 13 February 1993, pp. 33–37.

Landsborough, Diana, 'As close as your phone', *Reader's Digest*, September 1994, pp. 97–100.

Powell, Gareth, 'Fax, and nothing but the fax', *Sun-Herald* (Sydney), 24 October 1993, p. 44.

'The world's first fax machine', *New Scientist*, no. 1825, 13 June 1992, p. 11.

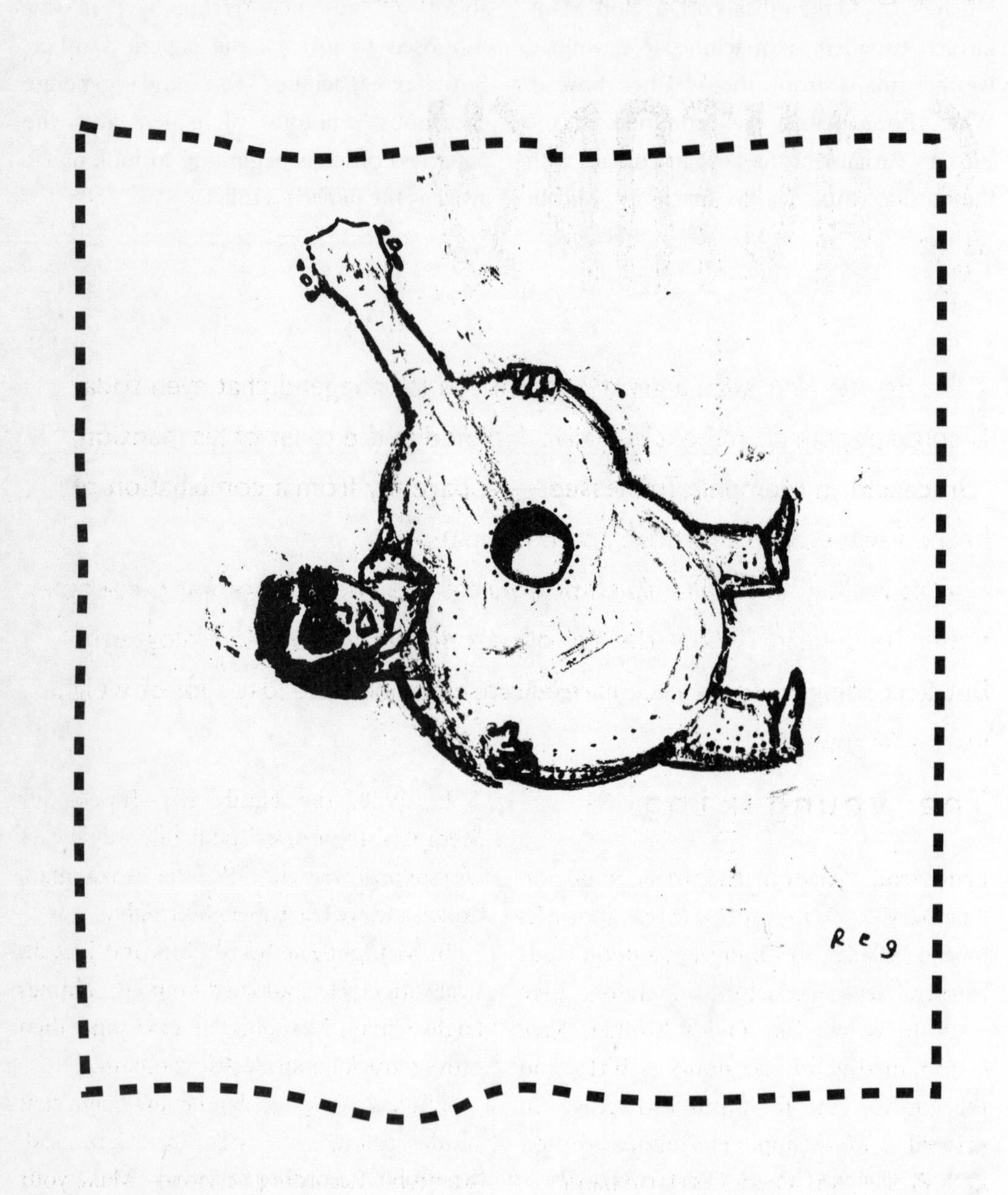

FAT DEAD ELVIS & AUTOPSY

Elvis Presley was such a great singer, and such a legend, that even today some people claim he is still alive. He died in the toilet of his mansion, Graceland, in Memphis, Tennessee — apparently from a combination of heart disease, drug overdose and constipation!

Elvis Presley, according to some sources, ate enough food for 60 people! When he died, in 1977 at the age of only 42, he weighed 159 kilograms. But, according to some recent research, he could have lost a lot of weight just by fidgeting!

The young king

Elvis Aron Presley was born at noon, on 8 January 1935, in Tupelo, Mississippi. His mother, Gladys Smith Presley, a poor, God-fearing, sewing-machine operator, gave birth to twins. But Elvis's brother, Jesse Garon, died within six hours of birth, and was buried in an unmarked grave in Priceville, Mississippi. His father, Vernon Elvis Presley, was a penniless farm hand.

Elvis was an only child, and grew up surrounded by gospel music. He sang Pentecostal Church music, heard black musicians singing the blues, and listened to 'Grand Ole Opry' on the radio.

In 1948, the family left Tupelo for Memphis, Tennessee. The family was always very poor. Elvis said: 'We were broke, man, broke, and we left Tupelo overnight.'

In Memphis, he heard blues and jazz on Beale Street. He graduated from L.C. Humes High School, Memphis, in 1953, and then drove a truck for an electric company.

While driving one day, he saw a sign that would change his life. The sign read: 'Memphis Recording Service — Make your own records. Four dollars for two songs.' Elvis wanted to record a song for his mother, as a birthday present. Memphis Recording Service happened to be the home of the record label Sun Records.

THE ELVIS SNEERING MUSCLE

Elvis was very popular as a performer. He would move his hips in a style that gave him the nickname 'Elvis the Pelvis'. When he appeared on television, his hip movements were thought to be so sexually suggestive that the cameras showed him only above the waist!

He also had a characteristic sneer. The muscle that curls the upper lip has the longest name of any muscle in the human body. It's called the *levator labii superioris alaeque nasi*!

Elvis met Sam Phillips, a rhythm and blues producer, who owned Sun Records. Sam had been previously known to say: 'If I could find a white man who had the Negro song, and the Negro feel, I could make a billion dollars.' Elvis was that man. He already had his magnificent voice, with its wide range and melancholy quality. Thanks to his childhood influences, his voice had that 'Negro feel'.

The Rock 'n' Roll legend began in August 1954, when WHBQ radio in Memphis played 'That's All Right, Mama'. Even though many stations refused to play the song because it was 'too black-sounding', enough people heard it, and liked it, to order 5,000 records in just a few days!

He released five singles, all of which sold well locally. They each had a country song on the A side, and a rhythm and blues song on the B side. Elvis had a series of gigs at the Grand Ole Opry, but they were not really successful. In November 1955, Sam Phillips sold Elvis's contract to RCA Victor for $35,000.

Elvis's star began to rise, and in July 1955, his record 'Baby, Let's Play House' hit the best-seller charts. In 1956, he released 'Heartbreak Hotel'. This was the first of his 45 records to sell over a million copies, and was the first record ever to reach Number One on three record charts simultaneously — the *Billboard* country and western chart, the rhythm and blues chart, and the pop chart.

He starred in his first movie, *Love Me Tender*, in 1956. He ultimately made 33 movies. The movies were cheap, easy and fast to make (they took only six weeks or so), and produced massive box-office returns. One MGM executive supposedly said: 'They don't need titles'.

In 1957, Elvis moved to his new home, Graceland. By the time he was drafted into the US Army in 1958, he had released 14 consecutive million-selling records.

He left the US Army in 1960. From then on, his singing style became softer and less vigorous. He made most of his movies during the 1960s and, in 1968, came back to television with an electrifying special simply entitled: *Elvis*. That same year, he began to tour with an orchestra and a gospel choir. His career began to rise again. He recorded 'In the Ghetto', 'Suspicious Minds' and 'Kentucky Rain', which were to become classics.

But he also began to suffer a personal decline. He now had a drug dependence (mostly barbiturates), and a middle-aged spread.

He died on 16 August 1977, at the age of 42. At the Baptist Memorial Hospital in Memphis, he was pronounced dead on arrival, of cardiac arrhythmia (erratic or irregular heart beat).

Peanut butter, bacon and strawberry jam

Elvis's diet must have had a lot to do with his tragically early death.

ELVIS'S RECORD RECORDS

Elvis Presley is the most popular recording artist in history. In August 1992, his estate was officially presented with 110 gold and platinum records.

Some 8 million of his records were sold in the five days after his death. This was a significant percentage of the 500 million records sold while he was alive.

Elvis had the most Top Forty hits (107), the most weeks at No. 1 position (80), the most Top Ten hits (38), and finally, the most consecutive No. 1 hits (10). He rightly deserves the title of 'The King of Rock 'n' Roll'.

The music writer Lester Bangs said: 'I can guarantee you one thing — we will never again agree on anything as we agreed on Elvis.'

The average diet depends on whether you are in that lucky one-third of the human race who live in a wealthy country, or not. Bangladesh is one of the dozen poorest countries in the world. Thirty million people (about half the population) live below the poverty line. These days, the poverty line is defined by food intake, rather than income. These 30 million people eat less than 7,000 kilojoules (kJ) every day.

The average person needs about 6,300 to 7,500 kJ per day as a bare minimum just to keep alive (assuming that he or she just lies in bed and does no exercise).

People doing very heavy work, such as Arctic or Antarctic explorers pulling sleds across frozen trackless wastes, need to eat about 42,000 kJ per day. Even then, they have difficulty in eating enough food to get their necessary 42,000 kJ per day. When Mike Stroud and Ranulph Fiennes crossed Antarctica on foot in 1992, they could eat only 23,000 kJ each day. That's about three times the intake of the average person, but they were working so hard in such a cold environment, they each lost about 23 kilograms in weight!

A BBC TV documentary, called *The Burger and the King*, deals with Elvis's diet. With Elvis, you can forget about your average 6,300 to 7,500 kJ per day, or even the 23,000 kJ per day of your average Antarctic explorer. Before he died, Elvis was eating about 420,000 kJ per day! That's more than enough to keep your average Asian elephant alive — or, enough to keep you or me alive for nearly 60 days!

Elvis had his favourite foods. They included fried peanut butter and banana sandwiches, lemon meringue pie, burnt bacon, and cornbread in buttermilk — all high-calorie foods with lots of fat. The

ELVIS'S JUMPSUIT RECORD

Some of Elvis's jumpsuits were very heavy, because of all the 'jewels' that they carried. Some weighed as much as 14 kilograms!

basic element of Elvis's daily food intake was a baguette some 30 centimetres long. It was stuffed with peanut butter, bacon and strawberry jam. Each such 'sandwich' had 176,000 kJ. In his final days, Elvis ate two of these 30-centimetre baguettes per day, together with midnight snacks of hamburgers and deep-fried white bread, as well as his favourite foods.

According to the BBC TV documentary, Elvis was addicted to food because of the poverty that he endured while growing up during the Depression in the 1930s. In the end, according to the TV documentary, his food killed him.

Fidgeting is healthy

One thing Elvis could have done, to lose a little weight, was to fidget!

Professor Leonard Storlein, from the Department of Biomedical Science at Wollongong University, was the one who accidentally discovered the weight loss benefits of fidgeting. He was measuring the total metabolic rates of humans. He used a device called a whole-room calorimeter. A whole-room calorimeter is just a small, sealed, room. Because the room was sealed, he could measure the total amount of oxygen his volunteers used, and how much carbon dioxide they made. From this, he could work out how much energy his volunteers needed just to keep alive, and how much extra they burnt up in moving around and fidgeting.

He found that this extra energy component varied from 840 kJ per day (for somebody who just sat around) to 5,000 kJ (for a dedicated fidgeter). The difference of 4,160 kJ is an amazingly large amount. According to Professor Storlein: 'A person would normally run 10 kilometres just to get rid of 1250 kJ.' So 4,200-or-so kJ is equivalent to a 33 kilometre run.

In other words, dedicated fidgeters, twiddling their thumbs, and bobbing up-and-down, and crossing-and-uncrossing their legs, can burn up as much energy as you would use in running 33 kilometres!

But even if Elvis had been into heavy-duty fidgeting, it would not have made much difference to his weight. Fidgeting could have chewed up 4,200 calories per day, leaving him with over 415,000-or-so calories still to burn up!

Autopsy

One thing we *do* know is that Elvis is really dead, despite what they say in the news. Elvis had an autopsy after he died. In an autopsy, the doctor removes and examines the heart, the brain, and various other essential internal organs. To get to these essential organs, the doctor has to open the body. In the words of the good doctor: 'If he wasn't daid before I did the autopsy, he sure was afterwards!'

There's an old saying in medicine: 'You can always make a correct diagnosis — but sometimes, you have to wait until the autopsy!' That old saying has a lot of truth in it. Even today, in Australia, the cause of death is misdiagnosed in about 50 per cent of elderly patients!

The autopsy is the one single test that will tell what actually killed the patient. The word 'autopsy' comes from the Greek words *auto* and *opsis*, and it literally means 'to see for oneself'.

The first recorded autopsies were carried out around 300 BC by doctors living in Alexandria. 500 years later, in 200 AD, medicine had advanced. The Greek doctor Galen actually compared what he found at autopsy with what he had seen on his patients, and what they had complained of.

The first known legal autopsy, to try to find the cause of death, was ordered by a magistrate in Bologna in 1302. Leonardo da Vinci and Michelangelo each performed autopsies, to understand the human anatomy better, and to improve their artistic skills. But the autopsy really became significant in 1761, when Giovanni Morganni published his great work *On the Seats and Causes of Diseases as Investigated by Anatomy*.

An autopsy first begins with a very careful examination of the outside of the body. Then, the front of the body is cut open, in the shape of the letter Y. Two cuts are made from near the two armpits, down to a point at the bottom of the sternum or breastbone, and then a single cut is made straight down to the pubic bone. The internal organs are examined, by the naked eye and with a microscope, the findings are recorded, the organs are replaced, and the cut is sewn up. Depending on what the pathologist is looking for, the whole procedure takes between two and four hours.

ELVIS WAS BURIED TWICE

Elvis was first buried in a mausoleum in Forest Hills Cemetery, in South West Memphis. There were several attempts to steal his body, so his family decided to move his remains. Afterwards, an ordinary piece of plywood was placed over the crypt, which was left with no identifying features. Nevertheless, his fans were able to find his first resting place, and placed flowers and mementos around the crypt. Since then, the plywood has been removed, and the crypt has been sealed with cement.

He was buried for the second, and last, time at Graceland. His name was misspelled on his grave! He was born with the name 'Aron', but the stone on his grave says 'Aaron'. Even more amazing, this error was never corrected. Some suspicious minds believe this to be proof that Elvis didn't really die!

Autopsies benefit us all

Even today, when we have access to all kinds of fancy internal diagnostic machines — such as computerised tomography (CT) scanners, magnetic resonance imaging (MRI) scanners, and even positron emission tomography (PET) scanners — there are many good reasons for performing autopsies.

For one thing, the only way we could tell that these scanners actually worked in the first place was by doing autopsies.

Autopsies are needed to see how prosthetic devices (such as hip joints and heart valves) actually last inside a body. Thanks to autopsies, the heart valves of the early 1960s were redesigned. Autopsies also told us that people who had heart transplants could, very rapidly, suffer severe disease in the coronary arteries.

Autopsies monitor new treatments. They told us that adriamycin, an anti-cancer drug, could cause heart failure — and as a result, the dosages of adriamycin were changed.

Autopsies find new environmental diseases. People who filled silos, and who did not wear breathing protection, suffered a lung disease now called 'silo-filler's disease'. Today, silo workers wear masks. And we've all heard of asbestosis, another disease discovered by the autopsy.

Autopsies are also the ultimate form of quality control. Even today, in the best teaching hospitals, major unexpected findings are uncovered at the autopsy. In about 10–15 per cent of patients, these findings would have led to a change in treatment and improved survival. In another 12–20 per cent of patients, these findings would not have led to a change in treatment and would not have changed their survival.

So it turns out that about 20–35 per cent of general patients die with a major

DID SCOTT OF THE ANTARCTIC STARVE TO DEATH?

Mike Stroud and Ranulph Fiennes each ate about 23,000 kJ per day, when they crossed Antarctica in November 1992, in temperatures as low as –50 Celsius degrees. Even so, they each lost about 23 kilograms in weight.

When Scott and his team crossed the Antarctic in 1912, they each ate about 18,000 kJ per day.

Mike Stroud — whose other job is at the Defence Research Agency Centre for Human Sciences — believes that 18,000 kJ simply wasn't enough. He calculates that the members of Scott's expedition would have each lost about 35 kilograms in weight! In fact, they probably starved to death.

diagnosis that was not found (before their death) by the hospital staff. This means that death certificates are wrong about 30 per cent of the time. Autopsies pick up the mistakes but, unfortunately, the rate of autopsies has dropped from 50 per cent about 50 years ago, to about 15 per cent today.

Autopsies are not very romantic, but we need them to plan the overall health care of Australia and to find new diseases.

This is what is written at the autopsy theatre at the University of Bologna, the home of the first legal autopsy: 'This place is where death rejoices to come to the aid of life.'

REFERENCES

Berlinger, Norman T., 'A mortal science', *Discover*, September 1994, pp. 30–35.

Geller, Stephen A., 'Autopsy', *Scientific American*, March 1983, pp. 110–121.

McKelvie, Penelope A., 'Medical certification of causes of death in an Australian metropolitan hospital', *Medical Journal of Australia*, vol. 158, 21 June 1993, pp. 816–821.

Mortimer, Derek, 'Fidgeters burn off most calories', *Australian Doctor Weekly*, 18 March 1994.

Mulley, Susan, 'Cause of death often incorrect', *Australian Doctor Weekly*, 17 February 1995, p. 48.

'100,000 calories diet brought Elvis unstuck', *Daily Telegraph Mirror* (Sydney), 25 December 1995, p. 13.

Seth, C., *et al.*, 'Diagnostic yield of the autopsy in a university hospital and a community hospital', *New England Journal of Medicine*, vol. 318, no. 19, 12 May 1988, pp. 1249–1254.

Spinney, Laura, 'Did Antarctic explorers starve to death?', *New Scientist*, no. 1996, 23 September 1995, p. 10.

Woof.
Gulp.
Reg

PYTHON GROWS GUT

The python is very different from other snakes. For example, some pythons in zoos have gone without a feed for up to two years!

When they're not eating, pythons shrink their gut (intestine) almost down to nothing — and when they are about to have a feed, they grow virtually a whole new gut! To grow that gut, they have to work harder than an Olympic 100-metre sprinter — and they have to keep it up for days, not just 10 seconds!

Snakes

Snakes have been around since the days of the dinosaurs. The oldest snake fossils go back about 120 million years. The dinosaurs died out some 65 million years ago, and by about 20 million to 30 million years ago, the snakes were really beginning to diversify into many different types.

Snakes are strange creatures. What other animal could survive with no external ears, no eyelids, no arms, no legs, and only one lung? Any other animal with these 'disadvantages' would be 'put out of its misery'. But snakes actually thrive — except where humans are around.

Snakes put some of their 'disadvantages', such as their long, skinny, limbless body, to good use. They can move very quietly, and are masters of camouflage. Their length and slenderness helps them get into almost any hiding place — either to catch food, or to hide away from predators.

Snakes range in size from the length of a child's hand (12 centimetres) to the length of a flatbed truck (10 metres). But it's actually a bit hard to work out these sizes accurately. Part of the problem is that during all their lives, snakes will keep growing longer. Their rate of growth will slow down as they get older, but they will still grow longer.

The smallest snakes, the thread snakes of the Middle East, keep well away from humans, and are really good at burrowing. This makes it hard to work out the shortest size of a fully grown thread snake. The longest snakes don't usually live long enough to reach their potential maximum size.

Snakes live on all continents, except Antarctica. In general, they prefer warm climates, and the greatest number of different snakes occurs near the Equator.

Snakes have not had very good press ever since that rather messy incident in the Garden of Eden with Adam and Eve. When people move into an area, snakes usually move out, or get killed. But they get their revenge — their departure is followed by a sudden increase in the so-called vermin population of rats and mice!

About a dozen cities in the USA have the annual attraction of the 'snake roundup', or 'snake rodeo'. Officially, the snakes that should be killed are usually only rattlesnakes and/or copperheads, but tens of thousands of other snakes are killed. It is to be hoped this dreadful custom will eventually wither away, but unfortunately, that will probably happen only after most of the snakes in a particular area have been killed.

Most people are scared of snakes, but most snakes are harmless. Australia is the only part of the world where poisonous snakes outnumber non-poisonous snakes. In fact, the world's most poisonous land snake, the Fierce Snake (*Oxyuranus microlepidotus*), lives in the western parts of Queensland.

Worldwide, about a million people get bitten by poisonous snakes each year. These million bites cause about 35,000 deaths each year. Most of these deaths happen in tropical countries, where there are many

SNAKES AND RELIGION

The snake appears in many religious beliefs around the world. The snake looks very different from every other animal, and can often deliver a swift and sudden death.

In some parts of the world, the snake is a symbol of something deeper — so the python of Delphi was a symbol of resurrection. Other peoples believed that even if they handled something as dangerous as a poisonous snake, their god would protect them. This is part of the rationale behind the Protestant sects in the USA who handle snakes.

The influence of the python was always very strong at Delphi. The priestess of the oracle at Delphi was called the Pythia. She would start proceedings by drinking water from a special spring, and would then enter a state of ecstasy. The python would then reveal sacred knowledge to her, which she would then utter in a very cryptic and enigmatic speech, which could be interpreted in many different ways. A priest would then help the seeker-of-knowledge to interpret these sacred words.

APOLLO KILLED PYTHON

Apollo was one of the best-loved gods. He was the son of Zeus (the King of the Gods) and Leto, and was the god of youth, beauty, light, and prophecy.

Python was a huge snake and was the daughter of Gaea (Earth). She guarded the oracle at Delphi, but actually lived a little out of town, in a cave on Mount Parnassus.

There are two different stories as to why Apollo killed Python.

One story says that Python had enjoyed a monopoly on giving oracles at Delphi, and Python did not want Apollo to start up another oracle in competition. So, for business reasons, Apollo killed Python.

The other story says that Python had been mean to Apollo's family, and he wanted revenge. Apparently she had persecuted Apollo's mother. So Apollo killed Python.

This killing of the fearsome Python was one of the early deeds of Apollo that helped make his reputation. He used one of his golden arrows in his silver bow. This was apparently the first time he had used one of his arrows in real battle.

After his victory, the priestess of the oracle at Delphi became his servant. Apollo wanted to celebrate his victory over Python with something big, so he called himself Pythius, and started the Pythian Games at Delphi. The winners of the various events (chariot races, feats of strength, and foot races), were honoured with crowns of laurel leaves.

Apollo was given many flowery titles by the Greeks. They called him 'destroyer of mice', 'destroyer of locusts', and even 'sender of fertilising dew'. Later, the Emperor Augustus promoted him to the position of one of the chief gods of Rome, and built a magnificent temple dedicated to him.

different types of poisonous snakes and, unfortunately, much poorer medical facilities. Death rates from snake bites are much fewer for the wealthy countries, which also tend to be further away from the Equator (and the snakes). There are about 1,000 bites in the USA each year, leading to about 15 deaths.

Python peculiarities

There are about 20 to 25 species of pythons. They live in the subtropical and tropical parts of Asia, Africa, Australia and the Pacific islands. They tend to be slow-moving and fairly docile. Pythons are not poisonous, and are supposed to kill their victims by stopping them from breathing. They then continue with swallowing their food.

Pythons are different from the other snakes.

Most pythons have two lungs. This is quite different from the other snakes, which have either a missing left lung or a left lung that is much smaller than the right lung.

Pythons also have two short spurs sticking out of their bodies, one on each side of their anus. These spurs are basically all that is left of the hind legs, which the other snakes no longer have. Pythons use these spurs in sexual intercourse. According to some scientists, these tiny hind legs show that pythons are still close, in an evolutionary sense, to their ancestors the lizards.

Mother pythons are probably the only snakes to have a maternal instinct. They hang around, after their eggs are laid, to turn themselves into a biological incubator and warm up the eggs. They do this by coiling about the eggs, shivering and contracting their muscles. This is very effective, and can warm up the eggs by 7 Celsius degrees. The eggs take about 70 days to hatch.

The general rule-of-thumb among herpetologists (reptile scientists) is that you should keep well out of the way of any python over 3.3 metres long. And keeping a 4.5-metre-long pet could prove fatal, as one American owner discovered. Attesting to the power of a large python, Jared Diamond wrote: 'When Secor brought his friendly 11-foot (3.3 metres) pet python Linus to our lab for a visit, I found that she could pull me in whatever direction she wanted.'

A hug before dinner

When a python goes for something small, like a rat, it will give its victim a quick non-poisonous bite to stun it. At the same time, the python will quickly wrap a few coils of its body around the rat. The python's teeth curve back into its mouth and dig into the rat's body. This means the teeth act like a one-way valve, so the rat can't pull itself out.

The current wisdom is that the python doesn't actually squeeze its victim to death, but suffocates it. Each time the rat breathes out, the rat makes its own ribcage a little bit smaller. At the exact moment that the rib-cage gets smaller, goes the current wisdom, the python tightens its grip another notch. After several such tightenings, the rat dies from suffocation.

This is not to say, however, that a python can't actually squeeze with a lot of force, as happened in Jalan Felda Redong, near

SNAKE SUPERLATIVES

The reticulated python (*Python reticulatus*) is a fairly slender snake, and ranges from the Philippines to southern Myanmar (Burma) to Indonesia.

According to the 1993 *Guinness Book of Records*, the record length for a snake is 9.99 metres, for a reticulated python shot in Celebes, in Indonesia, in 1912.

Again according to the 1993 *Guinness Book of Records*, a female reticulated python called Colossus held the records for the longest and heaviest snake ever retained in captivity. She died in Highland Park Zoo, Pennsylvania, USA, on 15 April 1963. She was 8.69 metres long, and when she was at her heaviest, weighed 145 kilograms.

Tenang, in the Malaysian state of Johor. In this case, a 6.65-metre python tried to swallow a rubber tapper, Ee Heng Chuan, aged 29. The python was found, in the act of swallowing the man head-first, on Monday 4 September 1995, at 10:30 pm . Ee Heng Chuan was already dead.

The python, when startled by the lights of the torches, released Ee Heng Chuan and then tried to slither away to safety. The police soon arrived, and shot it dead with M–16 rifles. The python weighed 140 kilograms and had a maximum diameter of 76 centimetres.

When Ee Heng Chuan was examined at autopsy, he was found to have multiple fractures in his neck and ribs. So the python was able to apply some muscular force of contraction — certainly enough to break the man's spine and ribs!

We humans can open our jaws some 30 degrees. A python has a specially hinged jaw, so it can open up to 130 degrees. In fact, the lower jaw has another hinge in it, held together by a strong ligament, which allows the jaws to open just a little bit wider again.

With a small meal like our rat, the python slides the rat in head-first, and then squeezes it down into its stomach.

If the meal has a large diameter, the python will use its very muscular neck and cheeks, in the same way we use our hands to pull on a rather tight sock. It will 'suck in' and push, first one side of the meal, and then the other side, in the same way that we pull on a tight sock first on one side, and then on the other side.

The python might, very rarely, have to spend a day or so swallowing a really long meal. So why doesn't it choke to death as the air supply to its two lungs gets interrupted? Because the windpipe of the python is both reinforced and mobile. This means that the windpipe can be safely shoved over to one side while the python continues to swallow and breathe at the same time!

Feast and famine

Now the python can go a long time between feeds — in some zoos, they've been clocked at going two years without a feed. It's common for a pregnant female python to go for 18 months without eating.

The main trick that the python uses to survive for a long time without food is to 'shrink' its gut. The gut, or intestine, is the tube that runs from the mouth to the anus. Along the way, there are long stretches of mucosa which are lined with tall, skinny cells. These cells are metabolically very active, which means that they use up a lot of energy just to keep them 'ticking over at idle'. So the python can save enormous amounts of energy just by making these cells 'shrink' away. The gut basically turns into a long, skinny, almost smooth, unlined tube.

But once the python gets food in its mouth, it can double or even triple the weight of its intestine overnight. In some experiments, scientists measured the rates at which the mucosa of a Burmese python and a sidewinder rattlesnake grew, in the first few days after a feed. They measured the rate of mucosal growth by measuring how rapidly amino acids (the building blocks of proteins) were absorbed into the mucosa. The normal rate of absorption increased by a factor of six after 24 hours, and by a factor of 16 after three days!

It's actually very important for the python to digest our rat as soon as possible

— sometimes, a python can die if it can't digest its meal immediately. The dead rat still has its own internal digestive juices, and its internal bacteria. These can combine to turn the python's meal — what should have been a nice little nutritious snack — into a rotting pulpy toxic mess. In some cases, the python has to vomit up its meal, so that it doesn't get food poisoning. Sometimes, the python can't vomit it up, and actually dies from food poisoning.

That's why the python has to be able to suddenly make huge quantities of digestive enzymes (60 times greater than normal), simply to digest this lump of incoming food.

If the python is able to bask in a convenient patch of warm sunlight, it can speed up its digestive process. (After all, most chemical reactions go faster at higher temperatures.) If the temperature is very low, as in winter, the python will simply not feed. Jared Diamond quotes the case of an Indian python trying to completely digest a rabbit at various temperatures. It took four or five days at 27.7 Celsius degrees, and seven days at 21.7 Celsius degrees. But at 17.7 Celsius degrees, the bunny was still not digested after two weeks!

The python has to work very hard indeed to grow so much new intestine. It uses up an enormous amount of energy to do this — as much as half of the energy in its potential meal!

You can get an idea of how hard it works when you look at how much air it has to move in and out of its lungs. When we humans run really rapidly, we increase the airflow in and out of our lungs by as much as 10 times. But suppose that a python has swallowed a meal weighing just 65 per cent of its weight. To grow the extra gut necessary to digest its meal, the snake has to increase its airflow by 36 times! And the python does this continuously for a few days or so, while lying on the ground motionless, appearing to have a gentle snooze! That lazy-looking snake is working harder than an Olympic sprinter.

While the python is eating and digesting, the amount of oxygen in its blood will drop to about 25 per cent of its

normal value. Even with that enormous increase in breathing rate, the oxygen level still drops by 75 per cent!

The python also has to generate enormous amounts of hydrochloric acid, simply to digest the meal. A balance has to be kept, and so the rest of its body becomes incredibly alkaline. If we humans were to become that alkaline, we would die immediately.

Keep Fido away from your python

A python can swallow something almost as big as itself. The following report, titled 'Snake Cannibalism', was submitted to *Nature* by H. Tsnagal, of Sancti Spiritue, in Cuba, over a century ago. 'While engaged in running a survey line for a railway across a wood in this district, I noticed a snake close to me, doing its best to get out of my way, but almost unable to do so. One of my men struck at its neck with his "macheti", and succeeded in cutting the snake's head clean off. Immediately, to our great surprise, another snake of the same species slowly emerged head first … A measuring tape showed that the larger snake was 6 feet in length and the smaller 5 feet.'

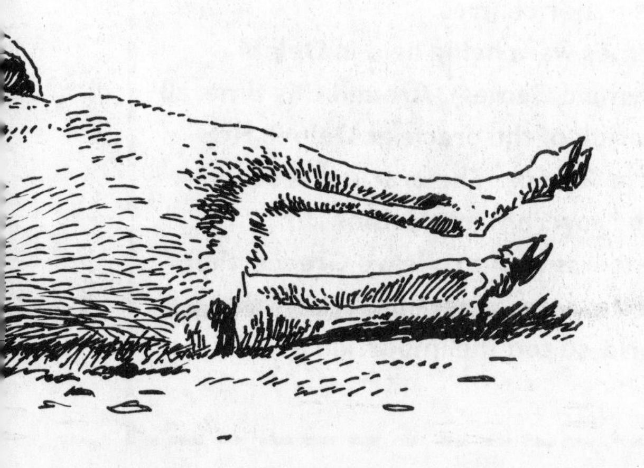

The python has the adaptations needed to survive an enormous meal.

The average large meal that we humans eat would weigh under 1 per cent of our body weight. But one viper easily beat this, by swallowing a lizard that weighed 160 per cent of its own body weight! Any python that's half trying can swallow 100 per cent of its body weight, while the average python will eat meals weighing about 65 per cent of its body weight. Practically all species of snakes can easily eat 25 per cent of their body weight at a single meal.

In medical schools, medical students are taught in terms of the 'average' 70-kilogram male. His average meal would be well under one kilogram. If he were able to eat like the average snake, he could easily eat a 17-kilogram meal. But if he were to copy the python, he could, with a bit of effort, eat a 70-kilogram meal — and that would include stretching his jaws to swallow something the size of a large pumpkin!

In one case, a rather large reticulated python swallowed a 14-year-old Malay boy on the island of Salebabu, in Indonesia. His body was mostly undigested, and easily recognisable, when cut out of the body of the python some two days later. In another case, an Asian reticulated python swallowed a 13-kilogram goat *and* an 18-kilogram goat at its first meal, and two days later finished off a 32-kilogram ibex. But the largest meal on record was a 59-kilogram antelope, swallowed by an African rock python.

Getting back to our rat, after two days its head is fairly well digested. By four days the chest and front legs have

TRADE IN SNAKES

There were about a million snakes imported into the USA over the two years of 1968 and 1969. These snakes would have sold for prices between US$5 and US$1,500 — so snakes are a multi-million dollar business in the USA!

THE FIRST SCIENTIFIC EXPERIMENT FOUND DELPHI

Zeus, the King of the Gods, apparently wanted to find the centre of the Earth. So he did the first recorded scientific experiment. He placed in position two eagles, one on each edge of the Earth (in those days, the Earth was flat, and had edges!).

The eagle on the west edge of the Earth and the eagle on the east edge of the Earth were released at the same instant. They flew at the same speed.

Where they met had to be the centre of the Earth. This place happened to be Delphi. Delphi is about 10 kilometres from the Gulf of Corinth, located on the lower, southern, steep, slopes of Mount Parnassus.

The exact point at which the eagles met is marked by a stone in the sacred temple of Delphi. This large stone, which is shaped like a conical beehive, is called the *omphalos*, which means 'navel'.

Delphi has been continuously inhabited since around the 14th century BC. But it really became a significant town around the 6th century BC, after the nearby town of Crisa had been destroyed in the Sacred War of 590 BC. Delphi rapidly became one of the region's major centres.

By 582 BC, the Panhellenic Pythian Games were being held in Delphi every four years (just like our modern Olympic Games). Around this time, all Greeks acknowledged the superior reputation of the oracle at Delphi. Her prophecies were supposed to be the best in Greece. The oracle was so powerful that she could even change major government decisions.

Every time that a new group of colonists was about to leave Greece, the oracle at Delphi gave them the benefit of its sacred utterances. So as the influence of Greece spread across the world, so too the importance and standing of the Delphic oracle spread.

TOP DOWN OR BOTTOM UP?

We humans control our gut in a different way from the python.

The activity of our human gut seems to be controlled from the bottom end. Suppose that a lot of undigested food arrives at our rectum, ready to be expelled into the outside world. In that case, signals are sent to our gut, to tell it to be more active. Our gut then becomes more metabolically active, and removes more of the nutrition out of the next meal while it's still in the top half of our gut. This means that less undigested food will arrive at the bottom end of our gut.

But the python gut seems to be controlled from the top end. Even as the meal is on the way down, the gut is growing in size and metabolic activity.

dissolved, while only an occasional bone remains after six days.

It can take up to two weeks for a rattlesnake to fully digest a rat, but bigger snakes with bigger meals can take several weeks. Practically everything is digested, except for some hair.

And this is what makes the python such an attractive pet. They're not very messy, they don't have to be fed often, and sometimes they'll even find their own food around your house. But make sure that you don't get one that grows too large, or one day, you might find your dog missing!

REFERENCES

Cossins, Andrew R. & Roberts, Neil, 'The gut in feast and famine', *Nature*, vol. 379, 4 January 1996, p. 23.

Diamond, Jared, 'The big squeeze', *Focus*, July 1995, pp. 20–24.

Encyclopaedia Britannica, 1996.

'Kuala Lumpur: python phobia', *Time*, 13 November 1995, p. 20.

'Python tries to swallow man', *Fortean Times*, February–March 1996, p. 12.

Tsnagal, H., 'Snake cannibalism', 100 Years Ago column, *Nature*, vol. 373, 2 February 1995, p. ix.

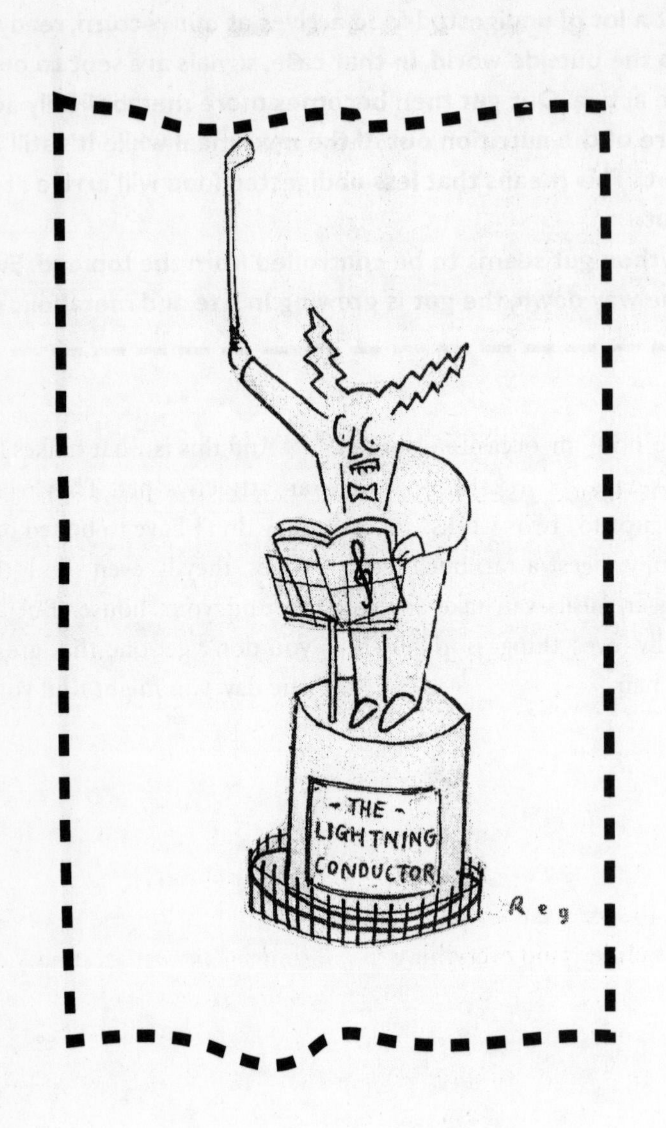
THE
LIGHTNING
CONDUCTOR
Reg

RED SPRITES & BLUE JETS

Lightning has always fascinated us. The Romans believed that Jove, the King of all the Gods, used lightning to punish people who did wrong. Jove also fooled around with lightning to just interfere with the normal course of human events down here on the ground. The 18th century statesman and scientist Benjamin Franklin was clever enough and lucky enough to run kites into thunder clouds without getting electrocuted.

Coloured lights

By the late 1980s, the scientists thought that even if they didn't understand everything to do with lightning, at least they had seen it all, and there probably weren't any major surprises left. But they didn't count on red sprites and blue jets — two of the amazing lightning phenomena that scientists have only recently discovered.

The first eye-witness anecdotes of rocket-like coloured lights above thunder clouds were made over a century ago. Back in the 1920s, airline pilots began to report that sometimes they saw something strange well above the thunder clouds, but nobody took much notice of such reports.

On the night of 6 July 1989, Richard Franz, Robert Nemzek and John Winkler from the University of Minnesota in Minneapolis were testing a new low-light-level black-and-white TV camera. Purely by accident, a thunderstorm was in the camera's field of view. When they played the video tape back, they noticed large luminous discharges above the thunderstorm. They also recorded a strange 'twin flash' that began at the top of a storm cloud, and then rapidly discharged upwards. This thunderstorm was about 250 kilometres away from them. The curvature of the Earth meant that they could not actually see the thunder clouds, but that they could see events above the

OTHER DIAGNOSTICS

Optical images tell us a lot about sprites and jets. But there are other diagnostic tools currently being used.

The optical spectrograph is used to tell us more about the colour of sprites and jets. When nitrogen is heated into a plasma, it gives off a reddish colour. Perhaps sprites are caused by reactions involving nitrogen?

Other tools used to examine the sprites and jets include radio devices, which look at Extra Low Frequency to High Frequency (ELF–HF) emissions from sprites and jets and their accompanying lightning strokes. Another interesting field involves Very Low Frequency (VLF) measurements of ionospheric heating effects that might be associated with the sprites and jets. Finally, continuous wave radar probes of sprites and jets can be used to work out the densities of electron clouds.

thunder clouds. Suddenly, the reports of the airline pilots took on a whole new significance.

On the night of 21 October 1989, astronauts on the Space Shuttle mission STS–34 recorded, using the shuttle's low-light-level TV cameras, their first sprite. They described it as 'our first vertical-like discharge moving out from the top of a thunderstorm that was being illuminated by intra-cloud lightning'.

Soon afterwards, other scientists at the Marshall Space Centre in Alabama wondered if these strange lights had been accidentally recorded on video tape previously shot from the Space Shuttle. When Bill Boeck, Otha Vaughan and Bernard Vonnegut (Kurt's brother) searched through old video tapes, they struck paydirt! They found many frames with 'glowing fingers' on them.

Red sprites

The red sprites are called 'sprites' for two reasons — they're very fleeting in their nature, and we really don't know what causes them. In the early days, red sprites had many different names, including 'cloud-to-stratosphere discharges', 'cloud-to-ionosphere discharges', 'upward discharges' and 'upward lightning'. Red sprites happen directly above active thunderstorms. They happen at the same time as cloud-to-ground and intra-cloud lightning strokes.

Sprites usually occur in clusters of two or three or more. Some of the very large sprites seem to be clusters of many individual sprites, which are packed together very tightly. Other sprites are packed together much more loosely. These sprite events can cover horizontal distances of over 50 kilometres, and can fill a volume of the upper atmosphere bigger than 10,000 cubic kilometres.

Sprites can be very powerful, with an optical power between 5 and 25 megawatts. (This is millions of times brighter than your 25 to 100 watt household light bulbs, which actually put out over 99 per

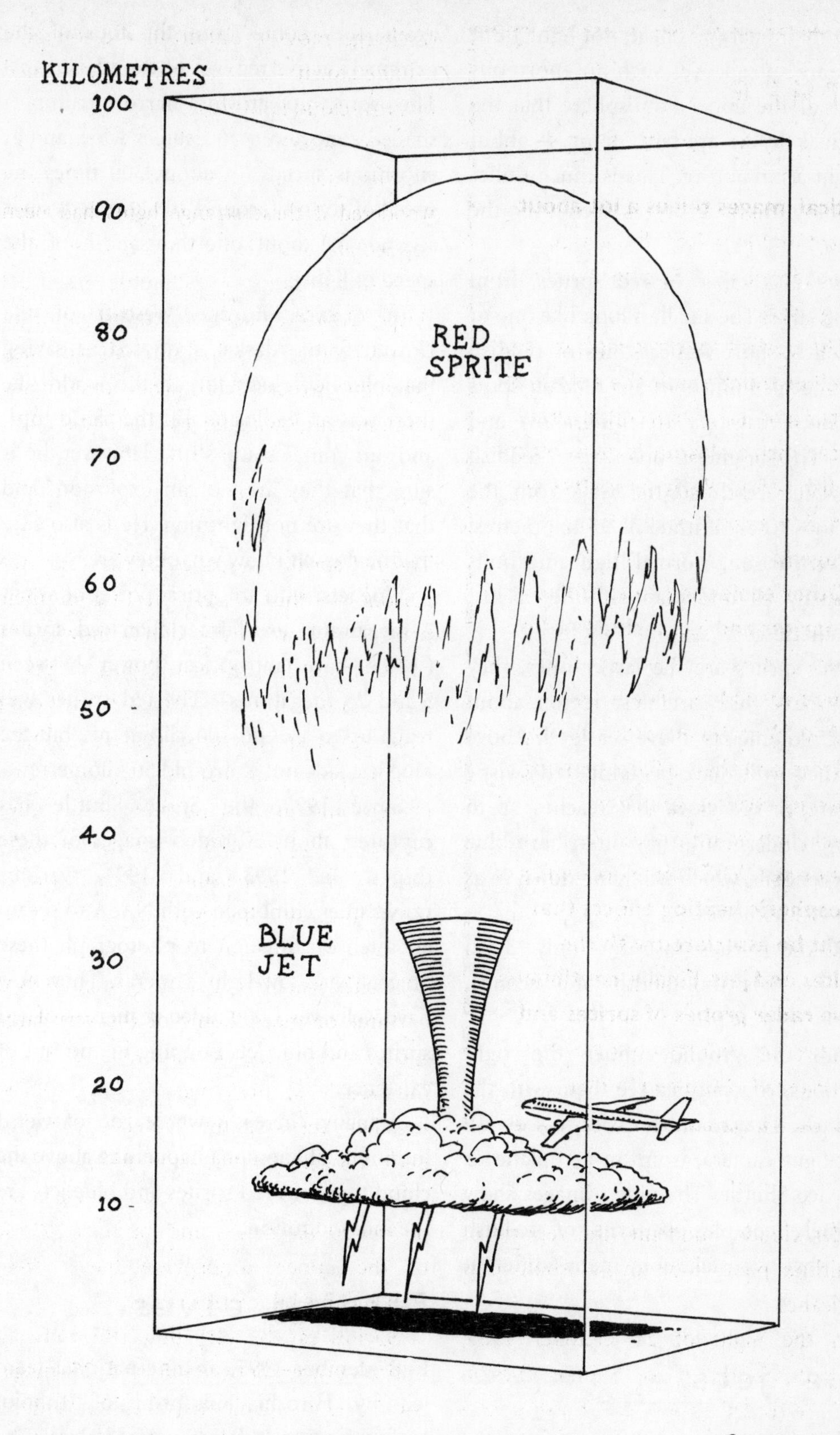
KILOMETRES
100
90
80
70
60
50
40
30
20
10
RED
SPRITE
BLUE
JET
©P. POUND 96

cent of their energy as heat, not light.) But sprites are spread over such an enormous volume of the upper atmosphere that the individual flashes are fairly faint — about as bright as an aurora. This is much duller than the glow of a city as seen over the horizon.

There is a range of red sprites, from small to large. The smallest look like one or more fairly short vertical lines of reddish light. The medium ones are reddish spots with faint extrusions stretching above and below. The brightest ones have reddish lines which reach all the way from the cloud tops to an altitude of 95 kilometres! (For comparison, 'normal' lightning bolts are usually all under an altitude of just 10 kilometres.)

If the sprites are the larger ones, they usually have their brightest region about 65 to 75 kilometres above sea level. Above this, they will have a faint red wispy structure, or red glow, that reaches up to the top. Underneath, they often have blue skinny tendrils, which will hang down to as low as 40 kilometres.

But the sprites last for only 3 to 10 milliseconds (thousandths of a second), and they're quite pale.

From the ground, under the right conditions, you can just see them with the naked eye. They're much easier to see with a low light camera, from an aeroplane or the Space Shuttle. The video images show amazingly complex patterns of reddish lights that can take on many different shapes.

Blue jets

The blue jets are quite different from the red sprites, and much more rare. Once again, they come from the tops of the extremely active regions of thunderstorms. They squirt upward in a narrow, trumpet-shaped cone about 15 degrees wide, and at enormous speeds — about 300 times the speed of sound (around 100 kilometres per second, or about one-thousandth of the speed of light)!

In *Science*, Eugene Wescott of the University of Alaska is quoted as saying that blue jets 'look for all the world like there was an explosion [in the cloud top], and out comes some stuff'. However, he is sure that they are not an explosion, and that they are not lightning. He is also sure that he doesn't know what they are.

Blue jets, with an optical power of about 2 megawatts, are fainter than red sprites (which have an optical power between 5 and 25 megawatts). The red sprites will reach up to around 100 kilometres, but the blue jets fade out at around 50 kilometres.

Since 1990, the Space Shuttle has captured about 20 video images of these things. In 1993 and 1994, various universities combined with NASA to set up an intense campaign to photograph these objects with low-light cameras. They now have well over 1,000 video sequences of red sprites and blue jets popping in and out of existence.

Actually, there's a whole zoo of weird luminous phenomena happening above the cloud tops, but red sprites and blue jets are the most common.

Finding elves

In December 1995, an international team led by Hiroshi Fukunishi of Tohoku University in Sendai discovered elves! The team was using a high-speed photometer near Fort Collins in Colorado to look at flashes of light happening over a thunderstorm that was situated on the Great Plains of the USA. That site on Yucca Ridge is so beautifully located that it can monitor thunderstorms up to 1,000 kilometres away. Elves are similar to red sprites, but are about 10 times shorter, and occur higher (70 to 105 kilometres) in the atmosphere.

Gamma rays

Scientists don't know what causes sprites and jets, or their characteristic colours. They also don't know why some sprites and jets seem to be associated with strange emissions of gamma rays.

In general, you get gamma rays emitted only when you're dealing with enormous energies, such as occur in nuclear reactions. But the orbiting Compton Gamma Ray Observatory has picked up a dozen or so gamma ray emissions coming from below

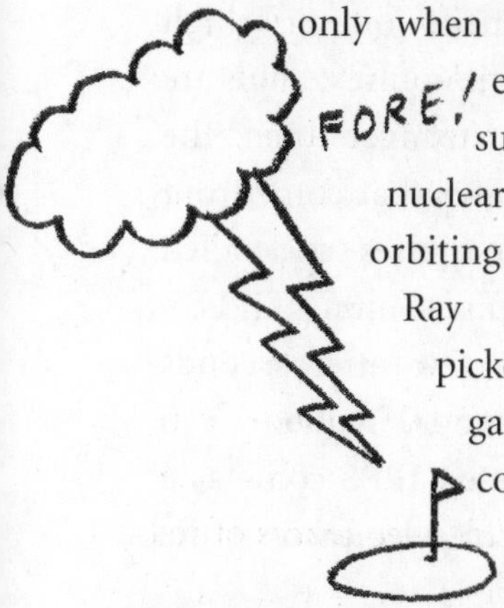

PEOPLE TO CONTACT, IF YOU SEE A SPRITE OR JET

Otha H. Vaughan, Jr, is the Principal Investigator of the Mesoscale Lightning Experiment (MLE). His snail-mail address is The Earth System Science Division, Space Sciences Lab, NASA/MSFC, Huntsville, AL 35812, USA. You can also contact him electronically, by way of email: skeet.vaughan@msfc.nasa.gov

Matt Heavner, at the Geophysical Institute of the University of Alaska, in Fairbanks, is also very interested in getting any reports. If you have access to the Internet, you can easily contact him via the dedicated 'Sprite Watch' home page on the World Wide Web at http://elf.gi.alaska.edu/sprites.html

LIGHTNING FACTS

Worldwide, about 1,000 lightning bolts hit our planet every minute.

Lightning can kill. In the USA alone, lightning kills about 100 people each year. That figure is greater than the number of people who are killed by tornadoes, or by flash floods.

When lightning hits a human, it often switches off the 'breathing pacemaker', but leaves the 'heartbeat pacemaker' still working. So, most times, all you have to do for a lightning victim is perform mouth-to-mouth resuscitation. After a few minutes, the person should recover.

LIGHTNING MYTHS

There are many myths about lightning. Lightning actually prefers to strike the same place twice — and usually for the same reason that it struck that place the first time. There's no truth to the myths that thunder can make unborn chicks die and cause milk to go sour.

Jove's favourite bird was the eagle, who would carry Jove's deadly thunderbolts to Earth in its claws.

Many native American societies thought that the 'thunderbird' caused thunder and lightning. Flashes of lightning would burst out of the thunderbird's eyes whenever it winked. The native Americans had a special interpretation for the peeled bark that you sometimes see after lightning has struck a tree. They saw it as a sign that the thunderbird had slashed the tree with its razor-sharp claws.

In parts of South Africa the locals also have a thunderbird, the 'Umpundulo', which, like the North American thunderbird, peeled bark from trees. But the Umpundulo made lightning not by winking its eyes, but by striking its bright feathers.

the Observatory — not from outer space. The gamma rays seem to be associated with thunder clouds. These gamma rays have to be emitted from a source higher than 30 kilometres, because at lower altitudes, the atmosphere would absorb the gamma rays, and they wouldn't reach the Observatory. These gamma rays last only for about a few milliseconds, and are quite powerful — greater than a million electron volts (MeV). The cause of these gamma rays is still a mystery.

Pulse pairs

Another mystery is the strange radio pulses that come in pairs. Again, these come from above thunderstorms, and again, they're probably associated with sprites and jets. They were first detected by the ALEXIS satellite launched in 1993 to monitor clandestine nuclear blasts via the very short radio pulses that nuclear blasts emit.

These strange radio pulses are extremely powerful pairs of very high frequency (VHF) radio pulses, and are about 10,000 times stronger than the other known radio pulses that come from thunderstorms. These pulses are called Trans-Ionospheric Pulse Pairs (TIPPS). They last for only a few microseconds (millionths of a second). Nobody can really explain why these TIPPS come as a pair of radio pulses, or why each of the radio pulses is so short.

There's a lot we don't know

All these reports (red sprites, blue jets, elves, gamma rays, TIPPS) suggest that thunderstorms might have a much greater influence on the middle and upper atmosphere than atmospheric scientists previously suspected. We have a lot to learn about the electrical environment of our upper atmosphere. Perhaps sprites and jets are essential to the working of the electrical circuits that run through, and around, our planet.

The US Air Force is definitely interested. They want to be sure that these strange lights (and other phenomena) are not dangerous to their planes or to the Space Shuttle.

Perhaps sprites and jets may create chemicals in the upper atmosphere that could have local or even global effects. For example, they may be involved in ozone production and depletion, or in the destruction of CFCs. At the moment, we simply don't know.

Some atmospheric scientists think that sprites or jets, or both, are a fundamental part of every, fairly large, thunderstorm system. So far, we don't even know much about where in the world they mostly happen.

What makes sprites and jets?

It's pretty hard to work out what causes them, as we still don't know exactly what they are. But we do know that they are not an aurora, lightning, or 'air glow'.

One theory is that, when there is a very powerful cloud-to-ground lightning bolt, the electric field surrounding the cloud changes more rapidly than usual. In fact, according to this theory, the electrical field increases so rapidly that the air above the cloud no longer works as an insulator. The electrical tension has to be relieved, and a giant spark bursts upward, to release the extra electric charge.

Another theory claims, once again, that the electric field changes very rapidly. This rapid change makes electrons, in the upper atmosphere, crash into nitrogen. The nitrogen then jumps to a higher energy state. As it falls down again, it gives off the red light that you see in a red sprite. However, this theory gives no explanation for the blue jets.

Other theories involve incoming cosmic rays knocking electrons up to higher energy levels. These energetic electrons supposedly then set off a chain of events that leads to sprites and jets. Another scientist thinks that the lights might be related to 'Sporadic E', a strange atmospheric phenomenon that causes strange radio reception.

There are so many theories that, according to John Molitoris from the Lawrence Livermore Laboratory: 'It's almost like a theory-of-the-week club.'

You can help

Don't think that looking for red sprites and blue jets is a job only for the professionals. After all, most of the discoveries of supernovas (exploding stars) and comets have been made by 'amateurs'.

Here's what to do. First, they happen only above thunderstorms. So you need a clear line of sight to see the region above the cloud tops. There should be no clouds blocking your view. If you're on the ground, your best viewing distance will be 200 to 300 kilometres away from the storm.

Second, red sprites are not very bright (even so, they are 2 to 10 times brighter than blue jets). They are visible only on a dark night. You definitely cannot see them at twilight. You want the region you're watching to be against a background of a very dark sky. If there's a full moon above the thunderstorm, its bright light will wash out your view, and you simply won't see the sprites, even if they are there.

Also, because the sprites not very bright, they can easily be 'washed out' by the light from the lightning strikes. If the thunderstorm is putting out too much light, you can block out your view of the thunderstorm with a sheet of cardboard, and you'll still see the sky above the thunderclouds.

If you're near city lights, you can't see sprites. Your eyes have to be fully adapted to the dark (anything between 15 and 45 minutes of avoiding bright lights).

If your eyes are adapted enough to see the Milky Way, they can probably see the sprites and jets. When your eyes are fully dark-adapted, you can see faint objects much better by not looking directly at them. So you will see sprites literally out of the corner of your eye.

Third, sprites are not very common. They happen with only 1 per cent of the lightning strikes that are of average brightness. However, they occur with 3 per cent of more powerful lightnight strikes.

You'll just be able to see the sprites as they flicker on for a few thousandths of a second. They happen much too fast to call them to the attention of somebody else to have a look at, but you can often see their strange dull red colour and vertically striated appearance. They cover about 10 to 20 degrees of the sky.

If you have the right kind of thunderstorm, and if you are in the right place, your chance of seeing sprites is greater than your chances of seeing meteorites (falling stars).

Walter Lyons, an atmospheric scientist from ASTeR, an atmospheric research corporation situated in Fort Collins, Colorado, described his first naked-eye sighting of a sprite as 'a faint, almost curtain-like gauze of salmon-coloured orange–red'.

And once you find them, report them to your local meteorological office in your nearest capital city — but quickly, before anyone steals your thunder!

REFERENCES

'Bursting in air', *Science*, vol. 264, 27 May 1994, p. 1233.

Fishman, G.J., *et al.*, 'Discovery of intense gamma-ray flashes of atmospheric origin', *Science*, vol. 264, 27 May 1994, pp. 1313–1316.

Hecht, Jeff, 'Tornado watchers get a flash of inspiration', *New Scientist*, no. 1998, 7 October 1995, p. 17.

Holmes, Bob, 'Airborne elves caught in the act', *New Scientist*, no. 2009–2010, 23–30 December 1995, p. 17.

Kerr, Richard A., 'Atmospheric scientists puzzle over high-altitude flashes', *Science*, vol. 264, 27 May 1994, pp. 1250–1251.

Kerr, Richard A., 'Stalking flashy beasts above the clouds', *Science*, vol. 265, 5 August 1994, p. 740.

Webb, A.F., 'Light on sprites', *New Scientist*, no. 1995, 16 September 1995, p. 53.

FADING FISHERIES

We humans have been catching fish for over 10,000 years. And all that time, the ocean has seemed big enough to give us all the fish that we could ever possibly want. But even though there's no lack of fish at the fish markets, and there are no long queues outside the fish shops, and there is no increase in the price of fish, it seems we are running out of fish.

Between 1990 and 1996 the worldwide fish catch levelled out to a maximum of about 100 million tonnes per year. During those six years, we tripled the numbers of fishing ships. These ships are not little aluminium boats with two people in them, but floating fishing factories the size of jumbo jets. And, although we're catching the same weight of fish, we're catching more smaller fish and fewer bigger fish.

Fish on our planet

About 70 per cent of the Earth is covered with water, so fish are found all over the planet, from Lake Titicaca in the Andes (some 3.8 kilometres above sea level), to the deep ocean trenches in the Pacific (some 7 kilometres below the surface).

There are some 20,000 different species of fish living in this water. In fact there are about as many species of fish as there are of all the other species of animals with backbones *combined*! Just for comparison, there are about 9,000 species of birds, 6,000 species of reptiles, 4,500 species of mammals and 2,500 species of amphibians.

About 60 per cent of these different species of fish live in the saltwater of the oceans, while the remaining 40 per cent live in freshwater.

EVOLUTION OF THE FISHES

Fish have been around for the last 510 million years. The first fishes, who were also the first creatures with a backbone, were the Ostracoderms. They appeared about 510 million years ago and were extinct by 350 million years ago. These fishes didn't have jaws, and they lived mainly in freshwater.

About 410 million years ago, the first fishes with jaws, the Acanthodians, or spiny sharks, appeared. They died out about 250 million years ago.

The modern bony fishes, similar to today's fish, first appeared around 395 million years ago.

Fish in our history

Fish have been a part of human culture for as long as we have had records.

Some 12,000 years ago, there were cultures that relied almost entirely on a diet of fish. The Maglemosian people in Scandinavia used fishing lines and nets, fishing spears that were tipped with pointed stones, fish-hooks made of stone, and harpoons made of bone.

About 4,000 years ago, an unknown artist made a bas-relief on the wall of an Egyptian tomb. It shows an Egyptian man angling for a tilapia fish in a fish pond in his garden. It also shows his wife unhooking two tilapia fish. This wall carving is the oldest picture that we have of 'aquaculture' (fish farmed in ponds).

Long before the time of Christ, the Japanese had laws about oyster-farming. In Ancient Greece, both Plato and Aristotle mentioned fishing; Aristotle also discussed oyster-farming. The Romans built special fish ponds so it would be easy to go out and hook a fish for supper. Pliny also discussed Roman oyster-cultivation.

Sporting anglers

Sport fishing, as we know it, began in 1496 when Dame Juliana Berners, the prioress of an Abbey near St Albans, wrote a book about the best way to use a fishing rod. Her book was called *A Treatyse of Fysshynge Wyth An Angle*.

Possibly the most famous fishing book ever written appeared in 1653 — Izaak Walton's book *The Compleat Angler*. He discussed more than just different ways of catching fish. He also commented on the life cycles and feeding habits of fish. He went further, and dealt with the intellectual and psychological struggle that exists between the angler and his or her quarry. Not only did he claim that it was a worthy thing to fish purely for the love of fishing, but he also claimed that to catch a fish using fair methods was a superior thing.

While Dame Juliana helped turn fishing into a science, Izaak Walton helped turn it into an art.

Brutally efficient inventions

Modern fishing vessels are brutally efficient inventions. They are usually powered by one or more diesel engines, and they are

THE LAW OF THE SEA

The legal side of fishing has always been important. If we are ever to have sustainable fishing in the open seas, we need to have a Law of the Sea that various nations are happy to abide by.

For most of our history, the Law of the Sea has always been laid down by a few of the most powerful seafaring states. In the distant past, we had the Shipping Code of Hammurabi (Babylon, 1750 BC), and then the Rhodian Sea Law (Byzantine Empire, 7th century AD).

In 1493, Portugal and Spain were the two dominant seagoing powers, so they 'divided' the oceans between them.

In the 17th century, Britain was the dominant maritime power, so it could propose whatever it liked as the Law of the Sea. It was John Selden who put forward the concept of the *mare clausum*. This 'closed sea' was owned by the local coastal states. If they wished, they could charge any tolls they liked from ships.

Later, the Dutch jurist Hugo Grotius put forward the different concept of *mare liberum*. This 'free sea' concept meant that the ocean was free, and open, to sea travellers of all nations.

On 6 December 1983, the representatives of some 119 nations gathered to sign the United Nations Convention on the Law of the Sea. It gave 40 per cent of the ocean near continents and islands to the states that controlled those continents and islands. The remaining 60 per cent of the ocean was granted the traditional 'Freedom of the Sea'. But the wealth of the deep ocean floor, which is some 42 per cent of the surface of our planet, was granted to the Common Heritage of Mankind.

But it's difficult to force fishing countries to act in the interest of the common good. For example, in 1991, it was obvious that swordfish catches in the Atlantic were dropping dramatically. There were calls to reduce the catch, to allow the swordfish to recover. The USA and Spain dropped their catches by 15 per cent. But Japan increased its catch by 70 per cent, Portugal's catch went up by 120 per cent, while Canada's catch soared by 300 per cent!

On 5 December 1995, a treaty to protect the fish stocks of the world was signed at the United Nations, in New York. It allows fishing vessels to be boarded in international waters if they are suspected of illegal fishing. But it could be a few years before it is fully ratified.

fitted with an incredible range of equipment to find and catch fish. The largest ships (mother ships) can even process and freeze the fish while at sea.

These ships usually use nets, or hooked lines, to catch the fish. The nets include purse seine nets (which are legal), or drift nets (which were banned in 1992).

A purse seine net is a long net that hangs down into the water like a curtain. It is suspended from floats. A boat tows it around a school of fish, so the fish become

BIG CATCHES

According to the 1993 *Guinness Book of Records*, the greatest fishing catch ever caught in a single throw was 2,768 tonnes. This was taken by the MS *Flomann* from Hareide in Norway, using a purse seine net in the Barents Sea, on 28 August 1986. More than 120 million fish were caught in this single catch!

The most valuable single catch was taken by the Icelandic trawler *Videy*, at Hull, in the United Kingdom, on 11 August 1987. The value of the 42,446 tonne catch was $473,957!

encircled by the net. Finally, the net is pulled shut at the bottom by a rope. The pulling of the rope is similar to the pulling of a drawstring in an old-fashioned purse.

A drift net is a nylon net some 15 metres high, and some 90 metres long, which floats just under the surface of the ocean. Nets can be linked together to be as long as you like. Fishing vessels would pay out many of these drift nets, and create an enormous floating wall that would drift through the ocean overnight. In the morning the vessels would reel in the drift nets and collect the catch.

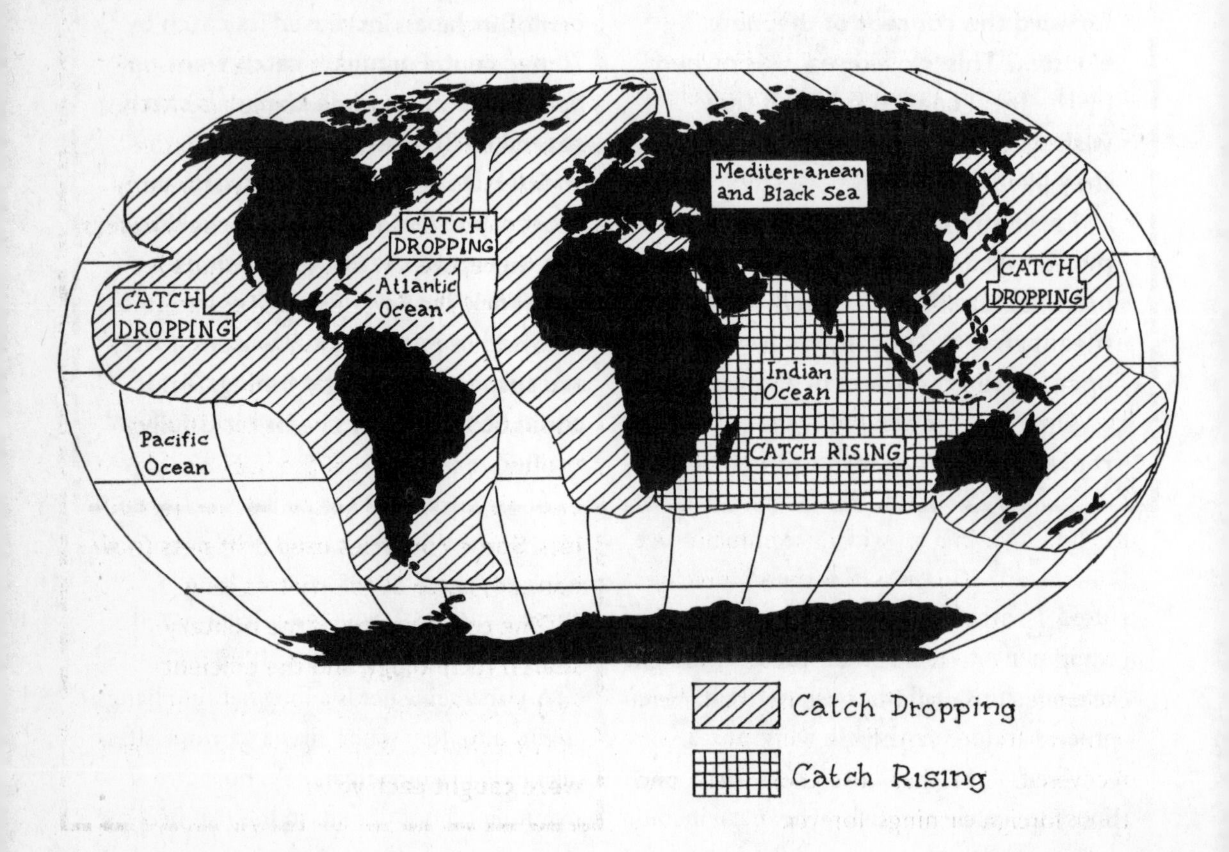

Great fisheries

In general, the best place to find fish is on the continental shelf — that underwater, fairly shallow, plateau that lies between the shoreline and the great ocean deeps. There is a lot of plant life for the fish to eat on the continental shelf, and the temperature, depth and currents are just right for the fish to live and breed.

It turns out that most of the great fisheries of the world's oceans are on the wider continental shelves (about 80 kilometres wide) in the mid to high latitudes, and usually at depths of 100 metres. These fisheries have been the major sources of fish for the great fishing fleets.

There are about 17 major great fisheries in the ocean — 13 of them have already collapsed, having been fished to exhaustion.

Only two fisheries currently enjoy increasing catches: the Western Indian Ocean fishery, and the East Indian Ocean fishery. This is because modern mechanised, heavily industrialised fishing has begun in those areas only recently. All the other fisheries in the world reached their peaks between 1974 and 1992.

In 1973, the great fishery off the coast of Peru supplied 22 per cent of all of the fish caught in the world. But most of the Peruvian fish catch did not get eaten directly by humans. Instead, it went into fish meal and fish oil, which were then fed to other animals, which eventually we humans ate. This was a waste of precious protein. For a short while, this fish trade provided one-third of Peru's foreign exchange earnings. But because the fishery was overfished, it collapsed. It has never recovered. Peru has lost those fish, and those foreign earnings, forever.

GREAT PREDICTIONS

The 19th-century naturalist Jean-Baptiste de Lamarck was wrong on a few of his ideas on evolution. He believed that characteristics that were acquired during a single generation could be easily passed on to following generations. So if you limped thanks to badly fitted shoes, then your children would also limp!

He was also wrong in his views on the immortality of marine fisheries. He wrote: 'Animals living in the sea waters are protected from the destruction of their species by man. Their multiplication is so rapid and their means of evading pursuit or traps are so great, that there is no likelihood of his being able to destroy the entire species of any of these animals.'

He didn't consider the adaptation of various military technologies. LORAN (Long-Range Navigation) enabled the trawlers to return to virtually the same spot, day after day. Radars let the trawlers get there, even in thick fog. Sonars let them find the fish, deep underwater.

The scale of the fishing gear is enormous. Longlines, baited with thousands of hooks, are up to 130 kilometres long. Some of the nets are big enough to swallow easily a dozen jumbo jets. Some countries used drift nets (now banned) up to 65 kilometres long.

The combination of the military search technology, and the efficient catching technology, meant that in some of the fisheries, 85 per cent of all the fish were caught each year!

IS AQUACULTURE GOOD?

In 1992, the marine biologist Dr Meryl Williams was made Chief of the Australian Institute of Marine Science. In 1994, she was made the Director of the International Centre for Living Aquatic Resource Management. This centre is one of 18 such research centres around the world that are looking at ways of better managing the oceans. Dr Williams thinks that part of the solution to our problems is to breed fish selectively, in the same way that for the last 5,000 years we have bred cattle and poultry.

She says that the tilapia, which is already being bred as an 'aquatic chicken' in more than 85 countries, is ideal for aquaculture. We've been using modern genetics and nutrition on our land animals for nearly a century. If we do the same with fish such as the tilapia, we can then begin modern fish-farming. She says that fish-farming is easy to integrate with agriculture on land — water from the fish ponds is rich in nutrients and can irrigate crops, and waste from the crops can feed the fish.

In 1983, the great crab fishery in the Bering Sea collapsed. It too has never recovered. In 1986, the last big scallop fishery in the Bass Strait went the same way.

The great fishery off the coast of Ecuador began to run out of fish in the early 1980s. It collapsed in 1987, with the onset of an El Nino. But when the El Nino went away, the fish did not come back. The great Ecuadorian fishery has never recovered.

The Grand Bank fishery, off the coast of Newfoundland, Canada, is 560 kilometres long, and between 37 and 100 metres deep. For a short time, it was one of the richest fisheries in the world. In 1968, they took out 810,000 tonnes of fish (1.4 per cent of the world catch for that year). But this dropped to 150,000 tonnes in 1977, and to 120,000 tonnes in 1991. The Grand Bank fishery collapsed in 1993, and fishing has been banned there since early 1995.

In California in 1995, the Coho salmon numbers dropped to 5 per cent of their previous level — again, thanks to overfishing. The same pattern has been repeated many times around the world.

Overfishing and collapsing fisheries

This massive fishing is not only virtually wiping out some species of fish, it's causing incredibly rapid evolution in the fish that remain. The remaining fish are now smaller!

Suppose that over a period of many years, you went to a cattle farm, and took away all the big cattle before they had a chance to breed. Only the shorter cattle would be left behind to breed. Over time, the average height of the cattle would gradually decrease. This has happened with fish all around the world, as the bigger fish have been taken away. For example, Canadian salmon, which have been very heavily fished, are now 30 per cent lighter than they were 30 years ago.

IS AQUACULTURE BAD?

In the last 10 years, the catch of fish from aquaculture has roughly doubled from 7.5 million to 15 million tonnes. In fact, aquaculture accounts for 20 per cent of all the fish actually eaten by people.

But aquaculture has meant that many mangroves have been destroyed to set up the aquaculture ponds. In fact, aquaculture is blamed for the destruction of half the world's mangroves! Many fish further down the food chain have been caught to feed to the fish in the aquaculture ponds!

Aquaculture can be bad for the local economy in poor countries. It needs large quantities of clean water (a rare commodity in many poor countries), and a large capital investment (which means that only the wealthy benefit). In many cases, aquaculture produces only expensive fish, which are then exported overseas. The local inhabitants have to then try to get fish out of the ocean.

Each of these fisheries that has collapsed has the same sad history.

First, there was a large stock of fish, and a small number of trawlers. The trawlers went in, and collected a large catch of fish — at least in the early years. Very soon, the number of ships increased, and the tonnage of fish caught decreased simultaneously. After a while, there weren't enough adult fish left to provide the next generation of baby fish, and the fishery collapsed. The fishing vessels then moved to another fishery. Nobody seems to have learnt from their mistakes.

Overfishing not only reduces the number of new baby fish that are born each

FISHING MAKES A LOSS!

There is no law that says that fishing has to make a profit. In the middle 1990s, the world catch was worth some A$70 billion. But it cost some A$124 billion to catch these fish — a loss of some A$54 billion!

This A$54 billion deficit is supported by various subsidies — price controls, very low interest loans, outright grants and fuel tax exemptions.

The fishing trawlers now have twice the fish-catching capacity of the annual catch (about 100 million tonnes per year). But the owners can't sell the trawlers, because they would make a loss. So they have to keep fishing, to repay their loans. In many cases, they apply political pressure to keep the catch limits and the subsidies high.

WE DON'T KNOW WHAT WE DON'T KNOW

The bluefin tuna is one of the most profitable fish to catch. But we simply don't know much about this valuable fish. We do know that its numbers have never been lower. The annual catch has dropped from 20,000 tonnes in 1970 to 2,500 tonnes in 1993.

Scientists from CSIRO Fisheries embedded tiny computer chips in the tails of bluefin tuna, and got a few big surprises.

First, the bluefin tuna swims right around the world, up to 15,000 nautical miles each year! Second, all the southern bluefin tuna come from one single stock, that have a *single* breeding ground off the coast of Indonesia. This means that if they are overfished, then they are gone. There is not another separate breeding ground from which the numbers can be built up again.

year, it also reduces the biodiversity in any area. The fishing fleets have had to move further down the food chain, to catch fish of lesser economic value.

In the 1980s, 30 per cent of weight of the entire world catch of fish was made up by only five species of fish. But this 30 per cent by weight of the world's entire catch accounted for only 6 per cent of its cash value!

Until around 1990, as our fishing technology improved, our catch increased.

World catch – five times increase in half a century

In 1948, the worldwide catch was 19 million tonnes of fish. In 1969, some 55 million tonnes of fish were taken from the sea. But 50 per cent of this catch was turned into products such as fish oil or fish meal, which were then ground up and fed to poultry and land animals, which were themselves later eaten.

In 1970, the fishery catch increased to 60 million tonnes; in 1978, to 70 million tonnes; in 1972, to 77 million tonnes. In 1986, the total fish catch for the world was 90 million tonnes, of which 40 million tonnes were taken by Asian countries.

But since 1990, even though we can still find fish at the fish markets, there have been worrying signs. It took 20 years, from 1970 to 1990, to increase the number of fishing vessels on our planet from 600,000 to 1.2 million.

From 1990 to 1996, we have tripled the number of fishing vessels so that we now have 3.5 million of them. But even though we tripled the number of fishing vessels, the world's fishing catch has remained virtually the same. In other words, to get the same amount of fish out of the ocean, we've had to fish three times harder!

At the moment, the total fishing catch for the world is around 100 million tonnes per year. About 15 million tonnes of this comes from aquaculture, while 85 million tonnes are taken from the ocean.

Of this 85 million tonnes, 20 million tonnes are turned into fish meal, while another 27 million tonnes are simply thrown back into the ocean, unwanted and dead!

Incredible waste

The polite word used to describe fish that are thrown back dead into the sea is 'bycatch' or 'bykill'. Worldwide, the average bycatch ratio is around four to one. This means that out of every 4 tonnes of fish that are caught, 1 tonne is simply thrown back overboard. That includes some 6 million sharks world-wide, each year!

But with crustaceans, the bycatch is much higher. Around 85 per cent of the weight of all the crustaceans caught in the sea is thrown away. This incredible wastage adds up to some 3 million tonnes each year. In the Bering Sea, in 1992, 16 million red king crabs were caught — and 13 million red king crabs were thrown back into the sea. In the Gulf of Carpentaria, the bycatch is around 10 to one. For each 25,000 tonnes of prawns that are taken out of the Gulf, 250,000 tonnes of fish are taken out, only to die and be thrown back into the Gulf.

There's another important issue with regard to prawns (and shrimps). The prawns move across a large area, but their breeding grounds are usually quite small. In the Gulf of Carpentaria, the breeding grounds for the prawns is the coastal fringes and mangrove swamps. These breeding grounds have a total area of around 1,000 square kilometres, which is smaller than Port Phillip Bay in Victoria!

A PRICEY TUNA

Fish can be worth a lot of money. In fact, the highest price for a fish was paid in Tokyo, in 1996: a bluefin tuna weighing 238 kilograms was sold for $105,000, which works out to $441.18 per kilogram.

International tensions

Because the fisheries have been collapsing, and because there were three times as many fishing vessels on the high seas in 1996 as there were in 1990, international tensions are rising.

PELAGIC VERSUS DEMERSAL

Two types of fish are taken in the saltwater fisheries — pelagic fish and demersal fish. The pelagic fish live near the surface — fish such as mackerel, sardines, herring, tuna, pilchards, anchovies and salmon. Demersal fish live near the bottom of the ocean — fish such as flounder, cod, sole, haddock and halibut.

Spanish and Portuguese fishing fleets have been working illegally in British and Irish waters.

On 9 March 1995, Canada took the unprecedented step of firing upon and seizing a Spanish trawler, the *Estai*, while it was in international waters. They claimed that it had been illegally fishing in the Grand Banks area, which had collapsed as a fishery. The *Estai* cut loose its net. But when the net was recovered from the ocean floor, it was found to have an illegally small mesh size. And in the hold of the Spanish trawler, the Canadians found thousands of dead cod.

Australia has had disagreements with five other nations over some 500 foreign ships that have illegally fished in our waters since 1985.

Britain has been sparring with Argentina over the toothfish. The toothfish is a large and tasty fish, up to 1.5 metres long, worth up to $8 per kilogram, and easily caught in the waters around South Georgia. The 'official' limit is 2,800 tonnes — total, for all ships, per year — but Britain claims that 'pirate' fishing trawlers have taken up to 15,000 tonnes.

Steve Blaber, a senior CSIRO Fisheries researcher, says: 'The good farmer looks after the land. He doesn't cut down the trees, or use up all the nutrients, or turn his farm into a desert. The same analogy applies to the sea. If you destroy the things the fish depend on, you create a marine desert.'

A very small number of dissenters claim that we don't really have a problem at all. In Scandinavia, records of herring catches go back some 500 years. In all of that time, there have been only four good periods of herring catches, each lasting about 50 to 80 years. The last one ran from 1885 to 1965. They claim that there really is no problem with the oceans, and that the collapsing fisheries are just part of a natural fluctuation.

On the other hand, it's quite clear that every fishery that has been heavily fished by humans has collapsed, and that the only two fisheries that have not collapsed have not — as yet — been heavily fished. By the year 2001 we will know the answer.

REFERENCES

Hickling, Charles F., 'The cultivation of Tilapia', *Scientific American*, May 1963, p. 143.

Holmes, Bob, 'Tough treaty to police global fisheries', *New Scientist*, no. 1989, 5 August 1995, p. 4.

MacKenzie, Debora, 'The cod that disappeared', *New Scientist*, no. 1995, 16 September 1995, pp. 24–29.

Parfit, Michael, 'Diminishing returns: exploiting the ocean's bounty', *National Geographic*, November 1995, pp. 2–37.

Safina, Carl, 'The world's imperiled fish', *Scientific American*, November 1995, pp. 30–37

Shepherd, John, 'Too few fish in the sea ...', *New Scientist*, no. 1996, 23 September 1995, pp. 48–49.

FORNICATION & FORGETTING

Our human memory is a very enigmatic thing. It's more than just something that can store and retrieve words and numbers. Our memory, that mysterious biological engine, is so powerful that it can recognise one person out of the 5 billion other people that live on our planet. Our memory is so large that it can hold a lifetime's experiences. And our memory is so lateral that it works on all of our senses. We can remember sounds and smells, tastes and feels, and emotions and pictures.

But this fantastic machine inside our skull can sometimes be unhinged by a single act of sexual intercourse. Sometimes if you fornicate, you forget!

HOW WE REMEMBER

There are three stages in the process of successfully remembering something.

First, you have to make a record of what you have experienced, and then put it somewhere in your brain. But sometimes a boxer can be hit on the head so hard that he will have no memory of anything that happened in the hour or so after he was hit. And yet, he can still continue to fight for several rounds, and even win the event. The boxer hasn't actually stored any new memories after the concussion to his head.

Second, you have to be able to store it for a reasonable length of time. Some alcoholics can store facts for only a few minutes, or sometimes, only a few seconds. Unfortunately, once they have forgotten the information, it's gone forever.

And third, to have a successful memory, you have to be able to pull that item of information out of your brain, and recall it.

Some memory!

Very few of us can use the full potential of our memory. But in February 1995, Hiroyuki Goto, 21, of Keio University in Tokyo, recited pi to 42,194 places (pi is the ratio of the circumference of a circle to its diameter). It took him a total time of 9 hours, 21 minutes and 30 seconds, including breaks totalling one hour, 26 minutes and 47 seconds. He beat the previous record of 40,000 places, set in March 1987, at the Tsukuba University Club House, by another Japanese man, Hideaki Tomoyori.

Most people can remember seven numbers in a row, but hardly any of us can remember 11 numbers. And so we humans use various memory aids.

The very first memory aid (that we know of) was invented in 489 BC, in Greece, by the poet Simonides. He had just finished reading some poetry in the home of Scopas, a wealthy nobleman. Simonides literally brought the house down, because, by an awful coincidence, the roof of Scopas' house collapsed just as the Simonides left. Unfortunately, all of his audience was killed instantly. In fact, the people were crushed so much, they couldn't be recognised. This was a major problem for the relatives, because they wanted to bury their own family members, but they couldn't recognise them.

Simonides suddenly realised that he had a visual memory of exactly where each person in his audience had been sitting. And that's how he invented the first memory aid — where you make a picture of each item of information you want to remember, and then you put it in a 'pigeon hole', next to all the other pigeon holes in your memory.

The best memory of any living creature probably belongs to a bird, the Clark's nutcracker, which lives in the American Southwest. This bird hoards food. Towards the end of summer, it harvests up to 33,000 pine seeds. Then it buries them in some 7,000 separate hidden treasure troves, each with about four or five seeds. The bird's memory is so good that it survives the winter, by returning to each of these individual 7,000 stockpiles, to dig up and

WHERE WE REMEMBER

In our brain, there are about 15 billion nerve cells — and each nerve cell can accept signals from, and send signals to, about 100,000 other nerve cells. It's a fantastically interwoven storage system.

Now the scientists don't really know what is going on, but they do think that they know a few of the areas involved in memory. They include the hippocampus, and the inferomedial parts of both temporal lobes.

eat the seeds! Only a few humans can approach this degree of memory skill.

Transient Global Amnesia

Now there have been a few cases of sexual intercourse interfering severely with a person's memory. The result is called Transient Global Amnesia (TGA). 'Transient' means for 'a short time', 'Global' means 'everything', and 'Amnesia' means 'loss of memory'. TGA was first described in 1964.

The medical textbooks say that most cases of TGA are caused by not enough blood flowing into certain parts of the brain. Those parts are currently believed to be the temporal lobes and/or the thalamus.

Sometimes, TGA can be related to an event such as sudden exposure to hot or cold water, a powerful emotional event, a headache, travelling in a car or, more rarely, sexual intercourse.

TGA usually affects men much more frequently than women. In an attack, the previously well victim suddenly suffers memory loss and becomes confused. There are usually no other symptoms.

In TGA, there are two simultaneous different types of memory loss.

One memory loss is called 'retrograde amnesia'. This type of memory loss reaches back to a period of time *before* the onset of the attack. This period of time of memory loss can range from a few minutes to a few hours, and (rarely) to a few years. As the TGA fades away, the length of the period of time of the memory loss gradually decreases. Memory of events *before* the TGA comes back, but not all the way. At most, the victim will lose about 45 minutes' worth of memories before the event.

SEX AND FAINTING

A very bizarre case (medical and legal) was that of a woman, in Cincinnati, who claimed that she had been sexually assaulted by a man who took advantage of her tendency to faint whenever she heard the word 'sex'. In fact, during the court hearing, the woman passed out twice, when the word 'sex' was spoken in the courtroom.

MEMORY AND MONEY

In Bogota, the capital of Colombia, some thieves have started using drugs to bring on a type of Transient Global Amnesia in their victims. Dr Alfredo Ardila of the Colombian Institute of Neuropsychiatry, and Carlos Moreno of the National University, studied 24 patients who had been admitted to the San Pedro Claver Hospital in Bogota with 'scopolamine intoxication'.

A typical case was that of a 28-year-old professional woman. She could remember that she left her office about 11 am, and that she was approached by a well-dressed man. Her next memory was waking up in the San Pedro Hospital some three and a half hours later. But during that missing three and a half hours she went to her office to get her salary cheque, went to her bank to cash her cheque, withdrew cash from her account, and went home to give away her jewellery to someone. But she had absolutely no memory of doing any of these actions. At the hospital, a urine check showed she had been given scopolamine. One of the other victims, however, did remember being sprayed in the face with something before temporarily losing his memory and getting robbed.

Scopolamine is that famous 'Truth Drug' of the black-and-white movies. It is also used to help with motion sickness. But this chemical from the datura plant has been used for centuries in rituals and medicines. It has been used as a sedative before surgery, and to calm down excited patients.

Usually, this period of time is much less — only a few minutes. (Retrograde amnesia can also happen in people suffering from delirium, alcoholism, and hypoglycaemia.)

The other memory loss is called 'anterograde amnesia', in which the victim temporarily loses the ability to lay down new memories. The sufferer knows who they are, and sometimes they know where they are. But they keep on asking what's going on. And when you tell them what's going on, they forget within a few minutes. The amnesia can last from a few minutes to a day. Anterograde amnesia can also be caused by benzodiazepams (e.g. Valium), alcohol, benzodiazepams combined with alcohol (e.g. on a long plane flight), and delirium in geriatric patients.

The TGA will gradually resolve over a variable period of time, ranging from minutes to hours. Finally, the victim is left with a memory loss. This memory loss is made up of two time periods — a period before the attack, and a period after the attack. Unfortunately, this loss of memory is usually permanent.

In some people, TGA can recur frequently. In some cases, a TGA is followed by a stroke. In some cases, a TGA can recur, but is not related to a stroke. In some cases, a TGA happens only once.

Confused in the USA

Dr Richard Mayeux, of the Neurological Institute in New York, reported one case of sexual intercourse causing TGA in a 64-year-old woman. She couldn't recognise her surroundings after she had sexual intercourse with her husband.

Dr Mayeux reported another case, involving a 47-year-old-man, where 'his wife found him in a state of confusion, wondering around the house, just after completing sexual intercourse ... The patient recovered within 24 hours, and has resumed his usual activities, including sexual relations with his wife.'

So while sexual intercourse can cause temporary amnesia, unfaithful partners shouldn't use it as an excuse, because the amnesia comes after the act, not before!

And all the time, you thought it was an orgasm!

Fornication and forgetting, Oz style

On one of my Science Talkback Shows on Triple J (Jenny Oldershaw, Thursday mornings, 11 to 12), a skydiving listener asked me a question about memory. 'How come', he asked, 'that sometimes a skydiver remembers

WORTH A TRIP TO THE LIBRARY

These articles are definitely worth a trip to your local large library. (The list of references at the end of this chapter gives the details of these articles.)

'Autoerotic Fatalities with Power Hydraulics', by R. L. O'Halloran and P. E. Dietz, discusses two men who became more than usually attached to their tractors — in specific, a back-hoe and a front-end loader shovel. One of these men had bought a tractor for himself as a Christmas present, and called it 'Stone'. According to the article: 'He wrote about Stone in a Christmas newsletter to his friends, in which he enclosed Stone's picture. He also wrote about his tractor in a long poem, which alluded to flying high in the sky with his friend Stone.'

Another interesting, if slightly gory, article is 'Temporary Ectopic Implantation of an Amputated Penis', by Hani S. Matloub *et al*. In this case, which has a happy ending, 'A 37-year-old man was doing repair work beneath his ride-on lawnmower when the motor engaged and resulted in avulsion [removal] of the penis. He was transferred to our institution, and immediately taken to the operating room.' His penis was in relatively good condition, but his groin, from where the penis originally came, was not. It was impossible to immediately connect his penis to his groin. So, as a temporary measure, it was stitched onto his forearm! (Is that what they mean by armed and dangerous?)

After a month, the swelling in his groin had resolved, so his penis was rejoined to its natural home. It is now functioning quite well, all things considered. He can urinate and experience ejaculations.

nothing of their flight, from jumping out of the plane to landing on the ground?' He said that this could happen even on a person's 300th jump.

The best explanation a few listeners and I could come up with was a theory that we invented on the spot, called The Overload Theory. This theory claimed (without any real proof) that sometimes, when the brain was overloaded with information, it would either refuse to accept any more information, or it would refuse to retrieve any new information. Either way, you don't remember the new information.

After the show, a gliding instructor rang in and said that he had occasionally seen the same loss of memory occur in glider pilots.

Then a woman rang in to tell me about her experiences with 'fornication and forgetting'. She had begun enjoying orgasms at the age of 11 all by herself, but had begun having them with her boyfriend at 17. She was now in her 30s, and eight-months pregnant with her first child.

She said that she would nearly always cry during orgasms, and could have as many orgasms as she wanted. In a way, masturbation was 'better' than sex with her husband, because she could control it better. With him, there was always the possibility that he would continue giving her orgasms after she didn't want any more.

She had always had a tendency to be 'away with the fairies' after her orgasms. She would usually continue sensual writhing and twitching for five minutes after she had stopped having sex. Even thinking about last night's sex would bring back many of the sensations of her orgasms.

She had TGA about 90 per cent of the time that she had sex. Her strongest TGA happened the second time that she had sex with her husband. She was completely disoriented for three hours, and she did not know who she was, where she was, or when it was. She said that as a result of this she 'hated having sex in the mornings'. But most of her TGAs lasted for much less than this. About 20 per cent of the time, they would last for about 30 minutes.

She had no idea why TGA happened to her, and to nobody else in her family — although she hadn't asked her brother. There was nothing remarkable in her medical or surgical history, nor in her family history, to suggest why she might have TGA after sex.

REFERENCES

Ardila, Alfredo & Moreno, Carlos, 'Scopolamine intoxication as a model of Transient Global Amnesia', *Brain and Cognition*, 1991, pp. 236–245.

Baringa, Marcia, 'Learning by diffusion: nitric oxide may spread memories', *Science*, 28 January 1994, p. 466.

Daily Telegraph Mirror (Sydney), '"Sex" knocked her out', 17 July 1993, p. 18.

Kosslyn, S.M., et al., 'Topographical representation of mental images in primary visual cortex', *Nature*, vol. 378, 30 November 1995, pp. 496–498.

Matloub, Hani S., et al., 'Temporary ectopic implantation of an amputated penis', *Plastic and Reconstructive Surgery*, vol. 93, no. 2, February 1994, pp. 408–412.

Mayeux, Richard, 'Sexual intercourse and Transient Global Amnesia', *New England Journal of Medicine*, 12 April 1979, p. 864.

Mishkin, Mortimer & Appenzeller, Tim, 'The anatomy of memory', *Scientific American*, June 1987, pp. 62–71.

O'Halloran, R.L., & Dietz, P.E., 'Auto-erotic fatalities with power hydraulics', *Journal of Forensic Science*, vol. 38, no. 2, March 1993, pp. 359–364.

Shanks, David, 'Remembrance of things unconscious', *New Scientist*, no. 1783, 24 August 1991, pp. 33–36.

Shettleworth, Sara J., 'Memory in food-hoarding birds', *Scientific American*, March 1983, pp. 86–94.

TRAFFIC JAM SHOCK WAVES

We've all had the fabulous fun of sitting in a traffic jam. Sometimes we're waiting to get out of the parking lot at our local church. Sometimes, we're stuck in 'gridlock' in a city at peak hour. ('Gridlock' is when all the vehicles in the grid of streets are locked into a frozen mass.) In these cases, you can actually see the other cars that stop you from moving.

But what about those weird traffic jams that you sometimes run into, out on the open road? In the middle of nowhere, you have to jam on the brakes, so you don't run into the car in front of you. Once you get going again, you keep an eye out for something that might have caused your sudden stop, but you see nothing! What's going on? The answer is … a shock wave!

WORLD'S BIGGEST TRAFFIC JAM 1

What would happen if all the cars went out onto the roads at the same time? Would there be any country in which it would cause a total traffic jam? The answer is no, and it's fairly easy to work out. All you have to do is divide the total length of the roads by the total length of the cars, in any particular country.

One of the most affected countries would be Italy. On average, each car would be crowded into just 12 metres of road. British cars would get 14 metres of road each. Even though Americans have the highest rate of car ownership (one car for every 1.3 people), they also have the best and biggest road network in the world. So each American car would get 33 metres of road. Australia has the same land area as the contiguous United States, but a much smaller population, so each Australian car would get 91 metres of road.

Ireland currently holds the 'car-freedom' record, with 111 metres of road per car.

WORLD'S BIGGEST TRAFFIC JAMS 2

According to the *Guinness Book of Records*, the longest traffic jam on record happened on 16 February 1980, in France. It was about 175 kilometres long, and stretched northwards from Lyons towards Paris.

Another enormous traffic jam happened on 12 April 1990. It was associated with the fall of the Berlin Wall, which had been built to stop easy access into West Berlin from East Germany.

The Wall came into existence on the night of 12–13 August 1961. By the 1980s, it had evolved into a complex system of fortifications, walls and electric fences up to 5 metres high, which were guarded by gun emplacements, watchtowers and dogs. The Wall itself was surrounded by cleared land that was laid with landmines. It extended some 165 kilometres through, and around, West Berlin.

The hard-line government of East Germany was forced from power in October 1989. One month later, on 9 November 1989, the new government opened the border between East and West Germany, which included West Berlin. The citizens of East and West Germany actually broke through sections of the Wall. Gradually, more and more people took advantage of the open border. The traffic reached a peak on 12 April 1990, when about 1.5 million cars tried to cross the border between East and West Germany!

Early roads

Paths and roads are one the first things you see when people move into an area.

The oldest known roads in the world are some 6,000 years old. They're a series of wooden walkways, in the southwest of England. They were built across a swamp to connect several villages. Eventually the wooden roads sank, and were beautifully preserved by the lack of oxygen in the swamp.

Building roads is one of the first things that a growing society does.

One of the earliest major roads was the Persian Royal Road. It stretched some 2,857 kilometres from the Persian Gulf to the Aegean Sea. This road was built around 3,500 BC and remained in use until 300 BC. Several short but very well-constructed ceremonial stone roads were built in the Middle East about 4,000 years ago. The Europeans built the 'Amber Routes', which stretched from the Baltic Sea to Greece and Tuscany. The Chinese road system, some 3,200 kilometres long, linked the major cities.

In South America, the Incas built a magnificent network of roads. By the 16th century, this network of two parallel roads (one in the Andes, the other on the coast) had many cross connections.

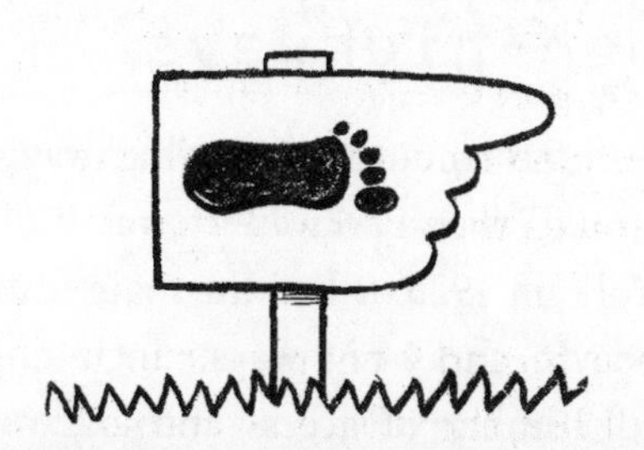

Roaming Romans

Around 2,000 years ago, the Romans realised how important good roads were for their military forces. At the peak of the Roman Empire, their network of roads stretched over some 85,000 kilometres. It sprawled from Britain to North Africa, and from the Persian Gulf to the Atlantic Ocean.

The Romans were the first road-builders to use good engineering practice, and their techniques were so good that nobody equalled their skill for 2,000 years. They started with earth or sand that was very tightly packed. Often, this was covered by a layer of large, flat stones. This was then covered with a layer of small pebbles, or gravel, that was set in mortar or mixed with lime. Finally, a thin, hard-wearing layer was laid on the top. The surface was curved so that water would run off to the sides of the road. (Sometimes, today's road builders forget to do this, so that puddles of water collect in the middle of the road!) Their roads were between 1 and 1.5 metres thick.

The Roman roads were so well designed that there were no real improvements for about 2,000 years. Even as recently as the early 19th century, Napoleon's engineers used Roman road-building techniques.

PATHWAY TO PRISON

Illegal marijuana growers can give themselves away by always using the same foot track to visit their 'plantation'. The police can sometimes see these tracks in aerial or even satellite photos.

EARLY TRAFFIC JAMS

As early as the 1st century BC, Rome had traffic jams. Julius Caesar passed a law that all vehicles (apart from chariots and government vehicles) were banned from the streets of Rome during the daytime hours. In the 2nd century AD, the Emperor Hadrian passed a law to limit the total number of carts that were allowed to enter Rome.

In the early 1500s, Leonardo da Vinci came up with a wonderfully advanced solution to the traffic jams that were happening in the busy streets of Italian cities. He put forward the idea of separating the pedestrians from the vehicles, by having them on different levels!

By the 17th century, several European cities had traffic jams so severe that laws were passed making some streets one way, and prohibiting parking on other streets.

Hit the tarmac

The next real improvement came in the early 19th century, when a Scottish engineer, John Loudon McAdam, invented 'macadamising'.

Like the Romans, McAdam started off with a foundation of tightly packed earth. The next layer included carefully sized, tightly packed stones, which were then overlaid by another layer of smaller, tightly packed stones. The gaps were filled in with small gravel, crushed pebbles and slag. This technique gave the road a stone surface that was resistant to water and the weather.

McAdam died in 1836. Eighteen years later, in 1854, E. P. Hooley, a Nottingham surveyor, noticed a strange thing. After someone had accidentally spilt a barrel of tar, Hooley observed that the tar turned into a smooth and hard surface. Tar is waterproof and flexible, so it will expand and contract as the temperature changes. Because of these advantages, tar was soon used as a covering layer for roads.

Hooley suggested that the new road surface should be named 'tarmacadam', after McAdam. Many people shortened the name 'tarmacadam', to 'tarmac'.

So that's how 'tarmac' was named after a person who had nothing to do with its invention!

Highways and super-highways

The very first modern 'super-highway' was the Bronx River Parkway. It was built in New York, in 1925. It had the main features of today's super-highways, including a limited number of access and exit roads, and the ability to carry a large volume of traffic at high speed.

Other countries soon followed — Italy with the autostrada, and Germany with the autobahn. An important consideration was the ability of these roads to carry military

WORLD'S FIRST MOTOR-VEHICLE ACCIDENT

In 1769, Nicholas-Joseph Cugnot built the first 'horse-less carriage' — and like many technological advances, it was funded by the military. His self-propelled road vehicle was an enormous three-wheeled, steam-powered, gun carriage. It travelled at only 1 kilometre per hour.

He built an even bigger one in 1770. It is supposed to have crashed into a wall in Paris, and so caused the first recorded motor-vehicle accident in the world.

traffic at speeds up to 160 kilometres per hour.

Probably the best highway network in the whole world is in the USA. It was built soon after World War II, and incorporated previous highways to make a network that linked all the major cities.

RECENT TRAFFIC JAMS

Until the 20th century, most people lived near where they worked — both employees and employers.

But now we have access to public transport (sometimes), and fairly cheap cars, we are more easily able to live and work where we want.

Part of the problem with public transport is that 80 per cent of the traffic load happens in just 20 hours of the week! This means that public transport systems need a large excess capacity — as far as average traffic volume is concerned.

Traffic flow – a strange beast

But no matter how well engineers design a highway, they can't design out 'shock waves'. One of the first people to think about shock waves was Robert Herman.

Back in 1963, General Motors in the USA had (believe it or not!) a Theoretical Physics Department, and Robert Herman was the boss. Today, many traffic engineers look upon Herman as the 'father of traffic science/traffic dynamics'. Even back then, Herman knew that traffic flow (the number of cars per lane per hour) is a strange beast.

Suppose that it's early morning, a few hours before peak hour. At this stage, there are only a few cars travelling along a single lane of road. The traffic flow per hour might be 30 cars. Gradually, as the Sun rises and more people head off to work, the number of cars in that lane increases, and so does the traffic flow. After a while, the traffic flow reaches about 2,200 cars per hour. As more cars enter this single lane of road, the traffic flow stops rising — it actually falls quite rapidly! It begins to break down into a 'stop–start' pattern, which is soon followed by a total traffic jam! Traffic engineers call this maximum traffic flow (usually around 2,200 vehicles per lane per hour) the 'capacity of the roadway'.

Traffic engineers had known for a long time that 'traffic flow' had this strange behaviour of gradually increasing until it reached a maximum of about 2,200 cars per lane per hour — but Robert Herman wanted to find out why.

Back in those days, they didn't have portable lasers or radar guns to measure distances and speeds in moving vehicles. So he joined two cars with a length of thin piano wire! To stop the wire dragging on the ground, or breaking, he added a motorised clutch on the back of the first car to keep the wire tight, as the cars moved closer together or further apart. Then he filled the boot of the first car with instruments to measure the distance between the cars and their relative speeds. He then said to the driver of the second car, 'Follow the lead car in what you consider to be a safe manner'. Then, he sent several pairs of these linked cars off to navigate the tunnels, roads and bridges of New York — including during peak hour!

NOISY ROLLER

One of the most famous slogans in the history of motor-vehicle advertising referred to the silence of the Rolls-Royce. It said: 'At 60 miles per hour, the loudest noise in this new Rolls-Royce comes from the electric clock.' The slogan was invented in 1958 by David Ogilvy. He was the founder of the US advertising agency Ogilvy and Ogilvy.

But the chief engineer of Rolls-Royce was not impressed by this snappy selling line. He shook his head sadly, and said: 'It's time we did something about that damned clock.' He got his wish in 1980, when the new Silver Spirit was fitted with a silent digital clock.

ROAD DEATHS

We humans are very illogical, especially when it comes to living with risks. One good example of this is the road toll.

Driving on the road is very dangerous, but most of us are hardly worried by it. However, we will get easily upset by low-risk situations — such as incredibly tiny amounts of a pesticide left behind on (say) the skin of apples.

In 1899, two people were killed in road accidents in the USA. In 1969, some 60,000 people were killed in road accidents in the USA. hose 60,000 fatalities in one single year were more than the American deaths suffered during the entire duration of the Vietnam War.

So far, in one century of motor vehicles, about 25 million people have been killed on the roads of the world.

Awareness distance, bunching and shockwaves

Herman made three discoveries during his traffic experiments.

First, the driver of the second car became 'aware' of the car in front at different distances, depending on the type of road. This 'awareness distance' meant that he would change his behaviour, depending on what the driver in front did. Obviously, if the car in front was 500 metres ahead, it would have no effect on him. On a multi-lane highway, with lots of opportunities for using other lanes, that awareness distance was about 55 metres. But in a tunnel, where the cars were confined to a single lane, the 'awareness distance' was about 75 metres.

Second, once the driver of the second car was aware of the first car, it showed a 'follow-the-leader' behaviour. The second driver tried to keep his road speed as close as possible to that of the first car.

Third, Herman also began to understand how shock waves can travel through the moving vehicles.

Bunching

Suppose you are in a helicopter, hovering above a single spot on the road. You can see the cars below, all roughly the same distance apart. Suddenly, you see that directly underneath you, the moving cars are all a little closer than average. They have bunched-up together. You still hover above the same spot. You then notice that this bunching-up of the cars is moving down the road, away from you. This bunching-up is a shock wave, that in this case is moving

THE FIRST TRAFFIC LIGHTS

The first traffic signal in the world was installed in December 1868, near the House of Commons in London, for the convenience of Members of Parliament. It was designed to control the flow of pedestrians, not vehicles, and it had two arms that a police officer could move up and down.

To make it useful at night, the engineers added some coloured lights. It was easy for the engineers to follow the system used on railroad signals: red to mean 'stop', and green to mean 'go'. The coloured lights were lit by gas, and were mounted on the top of a cast-iron pillar 7 metres high. The mechanism to change the colour of the light was linked into the mechanism for raising and lowering the arms.

In January 1869, the gas exploded. Unfortunately, the constable on duty was seriously injured. Even so, the traffic lights remained in operation until 1872, when they were removed.

The next set of traffic lights in Britain were installed in 1926, in London.

The honour for having the very first traffic lights for road traffic goes to the city of Cleveland, in Ohio. These traffic lights used red and green as the colours, and were installed in 1914.

down the road in the direction of the traffic. But shock waves can move against the direction of the traffic, or can even come to a complete halt.

Shock waves are usually caused by drivers who are not really aware of the cars around them. These drivers can either over-react (e.g., brake too hard) or under-react (e.g., are slow to notice that the cars in front of them are gradually slowing down).

Shock waves often happen in city traffic.

Suppose that you are driving in peak hour traffic, and all the cars are fairly close together. Suppose that the car behind you is a little closer than it really should be. Suddenly, the car in front of you brakes, and slows rapidly. Even if you are a bit slow in reacting, you will probably manage to stop in time, and not hit the car in front. And if the car behind is slow to notice that you have stopped, it will run into you! The car in front has not been hit, and can drive away, but your car is stuck to the car behind!

This particular shock wave has come to a complete halt, and on your rear bumper bar!

Shock waves also happen on the open road.

Imagine that there is a line of cars on the open road, all travelling at around 100 kilometres per hour.

The first car is driven by a driver who either reacts too vigorously or else has slow reflexes. He sees a rather handsome horse behind a fence, and brakes slightly to have a look. He slows down to 95 kilometres per hour, cruises at that speed for a hundred metres or so, and then gradually accelerates back to 100 kilometres per hour.

The second driver doesn't notice that the first driver has braked, until he is 50 metres

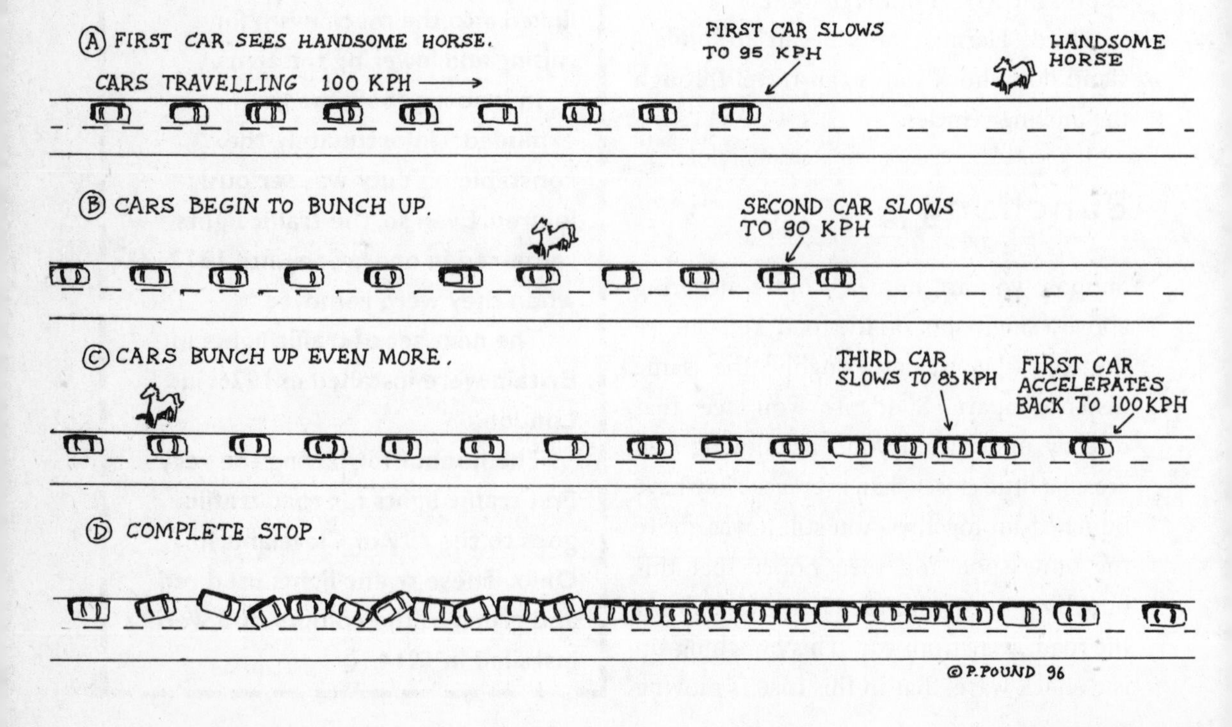

THE COLOUR-BLIND

Until as recently as the 1950s, many traffic lights were installed horizontally, rather than vertically. The design that we now use, with red at the top, was chosen to make it easier for colour-blind people.

By an unfortunate coincidence, the most common form of colour blindness is red–green colour blindness! That's why, in some countries, the red has a little orange mixed into it, and the green has some blue added, to make them a little easier to distinguish.

past the horse and already catching up with the first car. He brakes fairly hard, slowing down to 90 kilometres per hour. The third driver realises that he is about to run into the back of the second car when he is approximately 100 metres past the horse. He has to slow down to 85 kilometres per hour.

By the time we get to the 20th car in the line, the driver has to slam on the brakes to avoid a collision, and skids to a complete halt, about one kilometre down the road

GO FASTER BY GOING SLOWER!

The earliest computerised motor-vehicle traffic-control systems were installed in the early 1960s. Even today, the field of 'traffic dynamics' is still a new science.

A one-year pilot scheme was begun in 1995 on a 24-kilometre length of the M24 motorway, to the south-west of London. The road was fitted with electronic inductive loops under the road surface to measure traffic flow and vehicle speed. Large electronic display panels were installed over the motorway.

When the traffic flow reached 1,650 vehicles per lane per hour, the display panels showed a compulsory speed limit of 96 kilometres per hour. This dropped to 80 kilometres per hour at 2,000 vehicles per lane per hour. This relatively slow speed limit encouraged smooth driving, and reduced shock waves. It did this by discouraging drivers from weaving between lanes in an effort to find a faster-moving stream of traffic. The slower speed limit also reduced the number of drivers in the 'fast' lane. So all the lanes ended up carrying roughly the same amount of vehicles.

The results from the English study have not yet been analysed. But similar studies with 'variable speed limits' on the A5 autobahn, outside Frankfurt, showed that there were practically no shock waves, that accidents were cut by 29 per cent, and that traffic flow increased by 7 per cent!

HUMAN TRAFFIC JAMS

We all know about traffic jams in motor vehicles. But traffic jams also happen to humans on foot. One scientist turned a huge football stadium into one of the biggest laboratories in the world, and discovered that you can actually speed up human traffic flow by installing barriers!

Being in a crowd can change what you, as an individual, would normally do. Jean Cocteau, the French author and filmmaker, once wrote: 'If it has to choose who is to be crucified, the crowd will always save Barabbas.' And Virginia Woolf, the British novelist, wrote in her book *A Room of One's Own*: 'Great bodies of people are never responsible for what they do.'

Crowds can do terrible things, like riot, or lynch people. In April 1989, an awful crowd disaster happened in the United Kingdom at Hillsborough Stadium, when 96 people died in a crush after a gate was opened.

Certainly, dictatorships seem to love crowds. In Beijing in China, Tiananmen Square with its 40 hectares, or 400,000 square metres, can easily hold a million people. And Hitler planned to build a square in Berlin that could also hold a million people.

But crowds can also uplift the spirit. Many people love the feeling they get from being in a large crowd of fun-runners, like the 'City to Surf' crowd. Others have said that they were noticeably changed for the better by being a part of a religious crowd, such as in a Papal Visit or a Billy Graham Crusade. Every year, about 1.7 million Muslims do the pilgrimage to Mecca, to gather in the grounds of the Great Mosque. And the greatest crowd ever assembled occurred in 1989, to celebrate the Hindu Festival of Kumbh-mela, when 15 million people gathered at the junction of the Ganges and Yamuna Rivers.

Back in 1992, Keith Still was part of a huge crowd of some 70,000 people trying to get into Wembley Stadium for an Aids Awareness Concert. That's a big crowd. It weighs about 4,500 tonnes and, during a typical 90-minute concert, will breathe in 44 million litres of air. He was waiting to get into an entrance. He suddenly realised that the small part of the crowd that he could see was moving

from the handsome horse. The cars behind him have to stop, and they all ask themselves just how come they are stopped in the middle of nowhere. The first few cars, that actually set off the whole process, are still happily cruising down the road.

If you are hovering overhead in a helicopter, you can actually see this particular shock wave (the strange bunching-up of cars) begin at the horse, and travel one kilometre down the road, slowing down all the while, until it finishes in a bunch of stopped cars.

Herman found that about half of the drivers he tested set off shock waves either by over-reacting or by having slow reflexes.

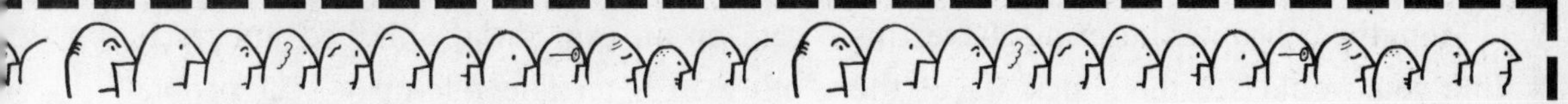

in a strange way. The people at the centre of the entrance, who supposedly had a lot of free space in front of them, were moving slowly, while the people at the sides of the entrance were moving much faster — exactly the opposite of what you might expect.

Keith Still started thinking about this, and tried to simulate the behaviour of crowds inside his home computer. But none of the simulations looked like real crowds. One day, at a cricket match, he noticed that 'each player took his cue from the bowler, and modified his stance accordingly'. In other words, the people in a crowd are not all independent, but react to each other.

He then came up with his two simple Rules of Crowds: (1) if the space ahead is free, move into it; and (2) if it's not free, then wait. Suddenly, his computer simulations began to work.

Back in 1994, he began analysing enormous crowds at Wembley Stadium, beginning with the FA Cup Final. Not only did he study the video tapes from 48 cameras around the stadium, he even perched on top of the entrance gates. He has now written a computer program that accurately simulates crowd behaviour — and it works for any crowd, whether it's there to see sport, popular music or ballet. For example, his program predicted that it would take 14 minutes and 30 seconds for a crowd of 20,000 to leave Wembley — and in real life, it took 15 minutes!

At entrance gates, most people head for the centre, and a traffic jam happens. Even though there's more room, nobody moves. But there are not so many people at the sides, even though there's less room, so the people move through faster.

A line of moving people gathers momentum, so sometimes repeatable patterns of behaviour happen in the crowd. Keith Still's computer program showed that if you place a long railing down the centre of an entrance gate, the crowd flow will go up by 25 per cent — and that's exactly what happened in real life when they installed the railings.

As our population increases, there will be more crowds — and computers will be there to help us!

And even if they didn't set off a shock wave, when one came past them, they would make it bigger, and would pass it on to the driver behind them! (Maybe it would be a good idea for people to have to learn about 'shock waves', in order to get a driver's licence.)

On the other hand, Herman found that every bus driver he tested was able to anticipate changes in the traffic around them. They could actually dampen down or even completely extinguish these shock waves, when the shock waves swept past them.

So suppose that about halfway in the line of travelling vehicles is a professional driver, like a bus driver or a truck driver.

From the helicopter hovering above, this time you would see the shock wave (the bunching-up of vehicles) gradually travel down the road until it hit the truck or bus. At that point, the shock wave would suddenly dissipate, as the truck or bus braked early, but not too hard.

It may seem odd, at first, that a shock wave can have an effect some distance away from where it first began. But consider a fireworks exhibition. You don't hear the bang when you reach the place where the fireworks exploded. You hear the bang when the shock wave reaches you.

All sorts of road conditions can cause these shock waves. An entry road onto an expressway can cause a shock wave, as can an exit road, and even a slight incline (if the driver doesn't realise that the car is gradually slowing).

Sometimes, these shock waves can have disastrous results. In France, during the summer vacation of 1982, a whole series of cars on an expressway smashed into each other in a terrible pile-up — 53 people died, 44 of whom were children.

But usually, the shock waves will affect only a few of the cars travelling. A typical such shock wave accident was reported in *The Telegraph* on 21 March 1996: 'Motorists who slowed down to look at one accident caused a second, equally serious, collision. An ambulance spokesman said both crashes occurred within 200 metres of each other on the Hume Highway near the Narellan Road off-ramp at Campbelltown between 8:05 am and 8:15 am yesterday. The spokesman said that four motorists — two from each accident — were trapped for about 10 minutes before being released by accident rescue officers.'

So if you keep your eyes on the cars in front of you, as well as behind you, you can stop those mysterious traffic jams in the middle of nowhere.

REFERENCES

'Accident causes collision', *Daily Telegraph* (Sydney), 21 March 1996, p 19.

Blair, Sean, 'The secret life of crowds', *Focus*, July 1996, pp. 27–29.

Cookson, Clive, 'Motorway snarls may be eased by go-slower scheme', *The Australian*, 5 September 1995, p. 43.

Herman, Robert & Gardels, Keith, 'Vehicular traffic flow', *Scientific American*, December 1963, pp. 35–43.

Holmes, Bob, 'When shock waves hit traffic', *New Scientist*, no. 1931, 25 June 1994, pp. 36–40.

Walker, Jearl, 'How to analyse the shock waves that sweep through expressway traffic', *Scientific American*, August 1989, pp. 84–87.

PHONE INVENTED 'HELLO'

Nowadays, we use the word 'hello' all the time, but linguists have always been puzzled about how it became so popular. Certainly, in the 1870s, the word 'hulloo' was used as a hunting call, while the London bobbies used 'allo, allo', as an expression of surprise, but the word 'hello' was not in general circulation at that time. And then suddenly, almost overnight, everybody was using this new word 'hello'. How come?

Professor Allen Koenigsberg, from Brooklyn College, claims that he has finally solved the famous 'Hello Puzzle'. After five years rummaging through the archives of American Telephone and Telegraph (AT&T), he says that the *telephone* brought the word 'hello' into our language!

Over 540 million phones on Planet Earth today

It's probably correct to say that the worldwide telephone system is the biggest machine ever built by the human race.

Nowadays, there are over 540 million telephones on our planet — roughly one for every 10 people. Phones carry voices, pictures and written messages. They do this by land lines (above and below the ground), sea lines on the ocean floor, microwave links and some 130 satellite links.

Tokyo has more phones than any other city in the world — a staggering 5.5 million phones. The largest switchboard in the world is at the Pentagon. It handles about

200,000 calls per day on some 25,000 lines, which have a total length of some 160,000 kilometres! Practically everybody in the wealthy countries uses a phone — except maybe the Trappist monks, who have taken a vow of silence.

Back in the 1700s, there was a phone equivalent of today's children's toy of tin-cans-joined-by-a-piece-of-string. But it was certainly not suitable as a reliable, long-distance, communication system!

But as engineers and scientists learnt more about electricity, many people tried to send voice by wire.

The most profitable patent

The first person to actually get a patent for a workable phone was Alexander Graham Bell. He signed it on 20 January 1876. However, he decided not to submit it to the American Patent Office immediately. But his father-in-law, Gardiner Green Hubbard, became impatient with waiting and submitted his son-in-law's patent on 14 February 1876, on Bell's 29th birthday. It was a little too close for comfort. Bell beat his competitor, Elisha Gray, by only a few hours! The designs of Bell and Gray were very similar — they each used the human voice to vary a resistance, which was then turned into variations in current.

On 7 March 1876, Alexander Graham Bell was awarded what was probably the most profitable patent ever issued — No. 174,465. It dealt with: 'The method of, and apparatus for, transmitting vocal or other sounds telegraphically … by causing electrical undulations, similar in form to the vibrations of the air accompanying the said vocal or other sounds.' (Actually, neither Bell nor Gray had built a working phone. This led to many legal battles, which Bell finally won. Bell's patent also described the concept of a complete telephone system, not just the telephone itself.)

Alexander Bell (he took on the name 'Graham' when he was 11 years old) was born in Scotland in 1847, the second of three sons.

UNDERTAKER INVENTED SWITCHBOARD

The first direct-dial exchange was installed in 1892 — because an undertaker was convinced that he was being robbed of rightful business!

Almon Brown Strowger was an undertaker in Kansas City. The wife of a rival undertaker worked on the local manual phone exchange. Over a period of time, Strowger became convinced that she was illegally diverting calls that were rightfully intended for Strowger to his competitor. He wanted a 'girl-less, cuss-less, out-of-order-less, wait-less telephone', so he invented a direct-dial telephone system, to bypass the operator. He set up the Automatic Electric Company in 1891.

His direct-dial system became essential. By 1951 there were so many phone calls made in the USA, that if they were carried through manually operated switchboards, half of the adults in America would have been needed to operate the exchanges!

THE RED TELEPHONE

In some movies, you will occasionally see, deep inside a very secure military centre, a red telephone on a table. This is supposedly the famous 'red telephone' that is a direct link between the White House and the Kremlin. They use the red phone only in times of great international tension. Unfortunately for the movie makers, the red phone isn't a red phone!

The red phone is actually white! Furthermore, it's not just a telephone link but a fax link as well. In fact, there's a double satellite link for voice, fax and teleprinter messages.

This 'hot line' was set up in 1963, immediately after the confrontation of the 1962 Cuban missile crisis. The very first message, sent on the 'hot line' on 30 August 1963, was: 'The quick brown fox jumps over the lazy dog. 1234567890.' As any typist in the English language knows, this sentence uses every key, and so it checks if all the keys on your keyboard are working. But the Soviets used the Cyrillic keyboard, and didn't realise the significance of this sentence and so their code-experts tried to decode this message!

In 1978, the 'hot line' was upgraded to run via a satellite link.

One occasion on which the 'hot line' was used to cool down a situation was when the Soviet troops invaded Afghanistan in 1979.

It's important that the line is reliable, so it has to be checked very regularly. But it's also important that no potentially inflammatory messages are sent on it. So 24 times each day, Washington and Moscow send messages to test the integrity of the line — and these messages have to be as neutral as possible. For example, on the turn of one hour, Moscow might send an encyclopaedia entry on women's hairstyles in 17th-century Russia, while in the next hour the White House would send excerpts from the rule book of the Professional Golfers Association!

His father, Alexander Melville Bell, specialised in teaching those who were deaf and who had difficulty in speaking. He invented a 'Visible Speech' alphabet, which showed deaf people just where to place their tongues in their mouths, and how to move their lips, so as to make the sounds of speech. His book, *The Standard Elocutionist*, went through nearly 200 editions! He had a position as Professor of Elocution at the University of Edinburgh, and brought up his children to follow in his footsteps.

Alexander Graham Bell's first job was teaching elocution and music to the students at Mr Skinner's school, in Elgin, in the County of Moray. By 1868, he was competent enough to assist his father, in London.

After Alexander Graham Bell's two brothers died of tuberculosis, his father moved the family in 1870 to a 'healthier' climate, Canada. By 1871, Alexander Graham Bell was teaching his father's speech system to the deaf in Boston. The next year he opened his own School for Deaf Mutes in Boston. In 1873, he was appointed Professor of Vocal Physiology at Boston University.

One of his deaf students was Mabel

TELEGRAPH VERSUS TELEPHONE

The telegraph was 'invented' in 1837 by Sir Charles Wheatstone and Sir William F. Cooke in Great Britain, and by Samuel F. B. Morse in the USA. It was operating commercially in the UK in May 1843, and in the USA in 1844. Just 30 years later, there were some 400,000 kilometres of telegraph lines in the USA.

A telegraph wire carries just a bunch of electric 'clicks', as a simple electrical connection is made and then broken. The timing of the 'clicks', via the Morse Code, makes up letters, which in turn makes up words. This is fairly slow, but it is robust. Even if the line is noisy, and the signal is weak, you can still recognise a 'click'. So a telegraph line can be quite long before you need to install a 'repeater' station to receive and retransmit the signal.

Telegraph lines crossed continents and oceans. The first successful transatlantic telegraph cable was laid between the USA and Europe in 1865.

But the situation is different with sending a voice signal down a wire. The frequencies involved are some 30 times greater in a telephone voice circuit than in a telegraph click circuit. After travelling through just 50 kilometres of wire, the voice signal becomes noisy, distorted, and unrecognisable. Many workers in the field felt that this would make it impossible to have long distance voice communication.

Technology came to the rescue, with three separate inventions.

The first invention was the use of 'hard-drawn' copper wire, instead of the steel or iron wire previously used. This did not attenuate, or weaken, the voice signal so much. The first test was successfully conducted between the cities of Boston and New York, in 1884. In 1891, the first telephone cable under the sea was laid between Calais and Dover. By 1893, the telephone line between Chicago and Boston was thought to be as long as telephone lines would ever get.

The second invention was the use of 'loading coils', or inductive coils. They also reduced the losses in the signal. The first test was in 1900, on a circuit in Boston some 400 kilometres long. This test was so successful that it was soon

Hubbard, whom he eventually married in 1877. She was 10 years younger than he. It was largely thanks to the financial backing of her father, Gardiner, and the father of another student, George Sanders, that he was able to continue his research. He continued to work on his multiple telegraph, and was granted a patent for it on 6 April 1875. But he soon became exhausted by his demanding work schedule, and had to stop for a while. In September 1875, he began to work on his voice telephone.

Alexander Graham Bell was incredibly talented. He was not very skilled with his hands, but he did have the ability to find and successfully work with artisans, like Thomas Watson, who could build exactly what was needed. He was a genius in turning his brilliant ideas into working machines.

followed up with a test on a line 1,000 kilometres long. Within 25 years, some 3 million kilometres of telephone circuits were loaded with about 1.25 million inductive coils.

The third invention was the amplifier, which was also called a repeater. In 1904, a primitive mechanical amplifier (with an electromagnetic receiver, and a transmitter made of carbon) was tried on a line joining Chicago and New York — but it was not really satisfactory. The electronic valve (or vacuum tube) amplifier was much better. The first telephone line across a continent was installed from San Francisco to New York in 1915.

By 1926, telephone conversations were able to cross oceans via shortwave radio links — but the quality was sometimes horrible. A three-minute phone call between New York and London cost US$75.

In the late 1940s, microwave radio links were introduced for overland telephone calls.

In 1956, the first transatlantic telephone cable was laid between Newfoundland and Scotland. This was possible, because of the invention of long-lived repeater amplifiers. This cable used 102 repeater amplifiers, and could transmit only 36 conversations at the same time. (Currently, the longest submarine cable is ANZCAN. It runs some 15,150 kilometres from Australia, to New Zealand, and thence to Canada.)

The concept of using satellites as relay stations was first proposed by Arthur C. Clarke in the October 1945 issue of *Wireless World*. He wrote that geostationary orbits (some 36,000 kilometres above the ground) were ideal, since a satellite in such an orbit would always stay above the same spot on the ground.

In 1958, the American satellite Score was placed in a Low Earth Orbit. Even though it had only a single voice channel, it could relay voice messages. In 1964, Intelsat 1 (Early Bird) was placed in a geostationary orbit above the Atlantic Ocean. It was powered by solar cells, and could transmit 240 conversations at the same time.

This is long distance

In 1880, he helped found the prestigious American magazine called *Science*. In the same year, France bestowed upon him the Volta Prize, which came with the sum of 50,000 francs (about US$10,000). He also invented the photophone, which sent sound on a beam of light, and the audiometer, which measures how good your hearing is.

In 1898, he followed his father-in-law, Gardiner Hubbard, in taking on the position of President of the National Geographic Society.

Alexander Graham Bell also worked on solar distillation, and sonar detection. He invented the aileron (a movable surface on the wing of a plane which makes it roll), and tricycle landing gear (which let planes take off and land on an aerodrome).

Way back in 1919, he even helped invent the hydrofoil speedboat that set a world water-speed record of 114 kilometres per hour, a record which stood for many years.

But when it came to inventing the correct etiquette for greeting people on the phone, he was a bit of a dill. After all, you speak to people on the phone differently from how you speak to them in person. In fact, the Japanese had to invent the phrase *mushi mushi*, for starting phone conversations.

All Bell could come up with was to suggest that people begin their telephone conversations with the word 'ahoy!', something he learnt as a child in Scotland.

Invention of 'hello'

It was Thomas Edison who probably pushed the word 'hello' into the general language. Besides inventing the electric light bulb, the film projector, and the carbon microphone (which was an essential part of the telephone for a century), he also invented the phonograph — the ancestor of the record player and the CD — which recorded sound on a cylinder of tin-foil and wax. He announced this

TRAPPIST MONKS

The Trappist monks were an offshoot of the Cistercian monks, who were founded in 1098 AD.

The Cistercian monks were named after the area in which they began, Citeaux, which has the Latin name of Cistercium. They were founded by a group of Benedictine monks who were unhappy with their easy life, and who wanted to live a stricter life. The Cistercians really began to expand their numbers in 1115 AD, when St Bernard of Clairvaux joined the order. By the time of St Bernard's death, he had personally set up 68 of the 338 Cistercian abbeys in existence. The Cistercians worked hard, and the abbeys owned large estates. So in Wales and Yorkshire, the Cistercians were largely responsible for the economic progress of the region.

But the wealth led to a softening of their hard ascetic life. After their Golden Age in the 12th century, the numbers of Cistercian monks dropped rapidly.

This was changed by Armand-Jean Le Bouthillier De Rancé. In 1664, he was made the abbot of the Cistercian abbey at La Trappe, in France. By 1678, the Pope had given him permission to set up a much stricter regimen at his abbey. He believed in absolute silence (apart from the chanting), hard manual labour and prayer, separation from the outside world, and a minimalist diet. He also thought that squalor, ugliness and poverty should be an essential part of life at the abbey.

Because the reform began in La Trappe, the monks were called Trappists. Their official name was the Order of the Reformed Cistercians of the Strict Observance (OCSO). Before the 1960s, they worked, ate and slept in continuous silence. But since then, following the modernisation of the Roman Catholic Church, many of the Trappist abbeys softened their rules.

So today some, but not all, of the Trappists observe the hard life of strict silence.

PHONES CATCH FISH!

In the state of Sabah in Borneo, 900 of the 3,500 pay phones were stolen in just one year. Local fishermen had stolen the phones, cut off the handsets, connected them to batteries, and lowered them into the water. The phones made a loud squealing noise, which attracted fish into the nets!

invention in 1877. In August of the same year, when primitive phone systems were just starting, Edison wrote a letter to a friend saying that probably the best etiquette for answering the telephone was to use an enthusiastic '*hello*', which was the exact word he had used after a particularly fruitful period of work on the phonograph.

Six months later, on 28 January 1878, the first commercial telephone exchange in the world opened at New Haven in Connecticut. At first, the operators were teenage men. But the young men didn't have the patience or courtesy that was needed, so they were soon replaced by women.

The official recommendation in their Telephone Operators Manual was that the operator should say either 'ahoy' or 'what is wanted?'. But the telephone operators preferred the short, sweet and snappy 'hello'. That first telephone exchange had only 21 subscribers, but within 10 years there were 200,000 subscribers in the USA, and within another 40 years, there were 30 million telephones in the world.

And as the phone exploded across the world, it took the word 'hello' with it.

In the 1970s, however, the official Australian telephone training manual for operators forbade them from mentioning the word 'hello'. Management had decided that if the operator said 'hello', the customer would reply 'hello', and then everybody would get very

UNTAPPABLE PHONES NOW TAPPABLE

Digital cellular phones were supposed to be introduced into Australia as part of the Global System for Mobile on 1 April 1993. Both of the networks involved, Optus and Telecom (later to be known as Telstra) proudly cited the security of the digital system. The digital signals were supposed to be encoded or encrypted before they were transmitted. This would make it virtually impossible for an outsider to tap into a digital phone call.

Unfortunately, various law enforcement agencies and the Australian Security Intelligence Organisation (ASIO) insisted that they should be technically able to tap into all phone calls. So, at the last minute, the introduction of the digital cellular phone network was delayed, so that the encryption software could be 'weakened'. After a suitable delay, digital cellular phones were introduced, and now our political masters and spy bosses can hear us get requests to pick up an extra litre of milk on the way home!

MORSE VERSUS QWERTY

The Morse code, which was used on the telegraph, was invented by Samuel Morse, who lived from 1791 to 1872. He checked with printers who, in those days, used individual blocks of lead — one for each letter. He worked out the popularity of the letters of the alphabet, by seeing how many of each letter they had in their tray of lead letters. He was sensible, and gave the shortest codes to the most frequently used letters. The letter E is the most frequently used letter in the English language, so the Morse code symbol for the letter E is the shortest code — a single dot.

Unfortunately, exactly the opposite happened with the typewriter! The early typists were fast, but the early typewriter mechanisms were clunky, and frequently jammed. So the keyboard was designed to slow down the typist. The commonest letters were placed away from the home keys (the ones directly under your fingers in the rest position), which is where the infrequently used keys (such as J) were put. That is how we ended up with the so-called QWERTY keyboard — QWERTY being the first letters of the top row.

And what is the only 10-letter word you can make using only the letters of the top row of a typewriter? Typewriter!

friendly, and waste precious telephone time!

Nevertheless, even though we've solved the mystery of how the word 'hello' came into our language, there's still another major telephone mystery left. Why is it that, on the telephone key pad, the numbers start with 1, 2, 3 at the top, while on every calculator, the numbers have 1, 2, 3 at the bottom?

Chronology of the phone

On 10 March 1876, three days after his patent had been granted, Alexander Graham Bell said the first complete voice sentence ever to be sent over wire: 'Mr Watson, come here. I want you.' In mid-October, Bell's wealthy future father-in-law, Gardiner Hubbard, 'borrowed' a telegraph line that ran between Cambridge Observatory and Boston, wired a phone to each end, and had the first-ever *long* telephone conversation, when he spoke for more than three hours with Thomas Watson.

In the same year, the Bell telephone patent was offered to William Orton, the president of Western Union, for $100,000. He declined the offer, calling it a 'toy', and a 'scientific curiosity'! (This patent led directly to the company, American Telephone and Telegraph. Just before it was broken up in 1983, AT&T was the largest company in the world, with over a million employees, and assets of £106,800 million!)

On 4 April 1877, Charles Williams of Somerville, Massachusetts, had the first-ever telephone line installed in a private house, when he had a line set up between his house and his office in Boston. The next month, on 17 May, the first telephone system in the world started up in Boston.

On 28 January 1878, in New Haven, Connecticut, the first commercial telephone exchange in the world opened with a total of 21 subscribers. It started off with day-time operation, but after six weeks, was running at night as well. On 21 February, the New Haven Telephone Company issued the first telephone directory, with 50 subscribers, on just a single sheet of paper!

Telephone numbers were invented in the next year, in 1879, during a measles epidemic. At that time, the telephone operators had to memorise the names of all of the people on their system. A doctor in Lowell, Massachusetts, was concerned that the telephone network would collapse if the operators became unwell. He recommended that numbers should be used, and that a list be written down so that the service could continue, if the original operators were sick.

On 3 June 1880, Alexander Graham Bell had the first wireless telephone conversation, on the photophone that he had invented.

By 1887, there were some 200,000 telephone listings in the USA. The first coin-operated telephone was patented in 1889, with 25 per cent of the take going to the inventor, 10 per cent to the store and 65 per cent to the local telephone company. There were about a million phones in the USA by 1900.

In 1915 to celebrate the first long-distance telephone service across the United States, between New York and San Francisco, Alexander Graham Bell in New York repeated his original 1876 sentence — 'Mr Watson, come here. I want you.' — to Thomas Watson, who was in San Francisco. But this time the call took 23 minutes to be connected and completed, and cost $20.70.

By 1929, there were some 30 million telephones in the world, with 20 million of them in the USA. The scientists at Bell Telephone Laboratories won several Nobel Prizes for their work. One Bell invention, the transistor, was first used commercially in 1951 in a direct-dial long-distance telephone system.

Tests began on cellular phones in 1978. The first cellular phone system began in Chicago in 1983, when motorists were offered cellular phone access for a connection fee of US$3,000, plus a service fee of US$150 per month.

Cellular phones were introduced into Australia in 1987, at the Sydney Opera House. I was the Master of Ceremonies for the event, as parachutists, canoeists and runners converged on the Opera House, all carrying their cellular phones. Unfortunately, Murphy's Law happened again, and the cellular phones could not be linked into the sound system for the attending journalists to hear!

Today, there are over 540 million telephones in the world. Americans use over 130 million telephones to make some 1,200 million phone calls each day!

REFERENCES

Bremner, Charles, 'Phoney ring to Bell's "ahoy"!', *The Australian*, 7–8 March 1992, p. 4.

Dunlevy, Lyn, 'ASIO link in delay to "untappable" mobile calls', *Sydney Morning Herald*, 31 March 1993, p. 7.

Landsborough, Diana, 'As close as your phone', *Reader's Digest*, September 1994, pp. 97–100.

'Phoning up for fish bait', *Daily Telegraph Mirror* (Sydney), 26 April 1996, p. 29.

'Sounding red alert', *Did You Know?*, Reader's Digest (Australia), Sydney, 1991, p. 187.

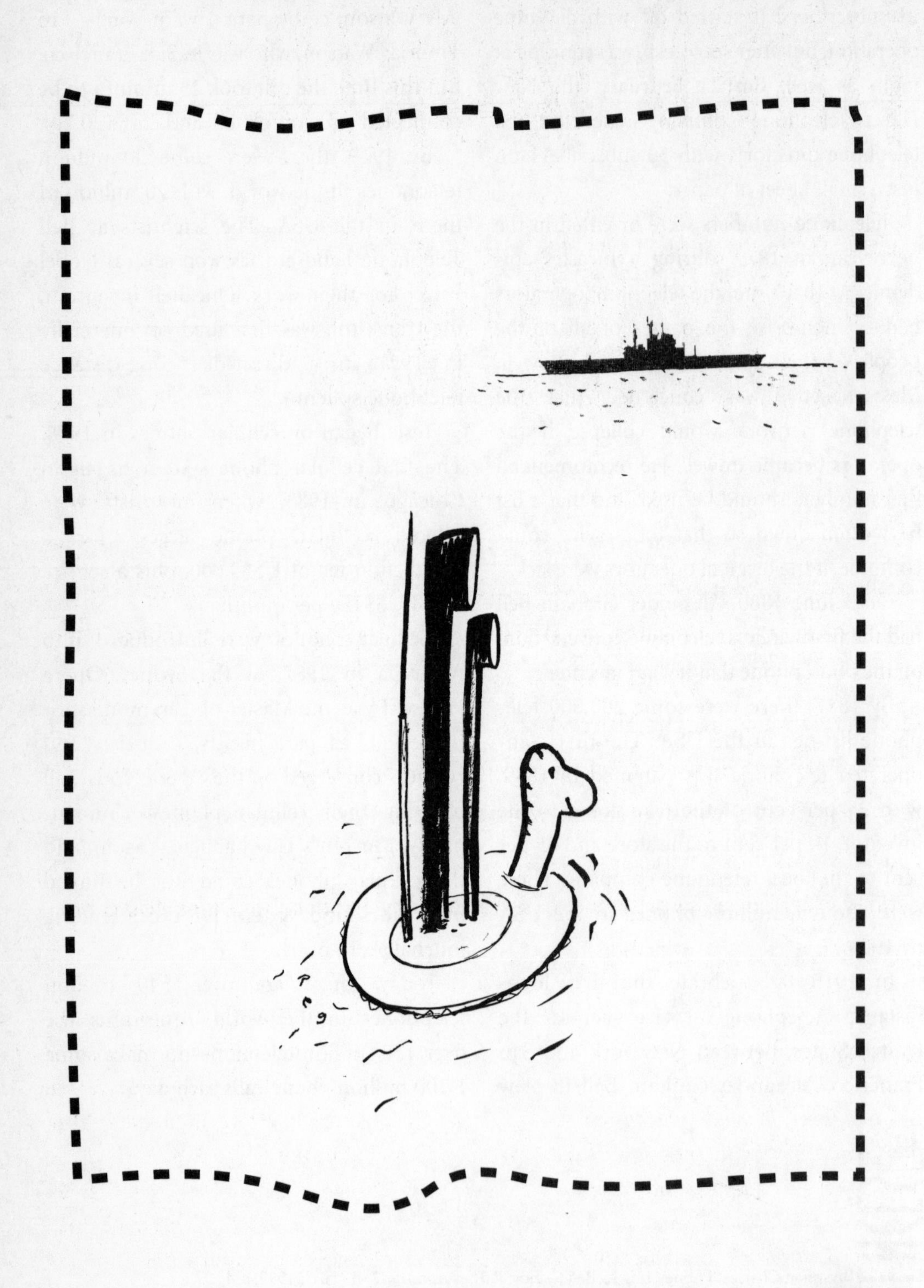

THIRD WORLD SUBMARINES

We humans can only walk on the land, so it's no surprise that we dream of flying in the air and swimming underwater. Our past mythologies gave us the legend of Icarus flying too close to the Sun, and of Alexander the Great cruising underwater in his glass barrel. But now it seems that in the near future, the desire to travel underwater could lead to strange new Cold Wars amongst the Third World countries.

Submarines were devastating in both world wars. Submarines almost brought England to its knees in both wars, because they sank some 32 million tonnes of shipping. Japanese submarines even turned up in Sydney Harbour in World War II. And one day soon, submarines could be devastating tools in terrible new wars.

The first submarines

People have wanted something like a submarine for thousands of years. The underwater glass barrel of Alexander the Great was probably a legend. History claims that the courier of Lucullus (the Roman general who fought Mithradates VI from 74 to 66 BC) really was able to use a primitive underwater apparatus (a goatskin!) to get a message through the enemy fleet.

It does seem that in 1150, during the Crusades, when the Christians blockaded Ptolemais, one of the enemy managed to get through the blockade in some kind of underwater apparatus.

The 12th-century German poem 'Salman and Morolf' describes an underwater boat that uses leather in its construction.

In 1578, William Bourne, an English mathematician, wrote a book called *Inventions and Devises*. His '*18th Devise*' was a submarine. It was a rowboat that had a wooden frame totally covered in greased leather, and which could travel underwater. It would submerge by becoming smaller! The sides would come closer together by way of a mechanical arrangement, and the submarine would slowly sink beneath the surface.

Twelve-man sub of 1620

The earliest documented submarine was built in 1620, by the Dutchman Cornelius Jacobszoon Van Drebbel. Drebbel had emigrated to England in 1604, and soon had James I as his official patron. Over the next four years he made many test-runs on, and under, the River Thames.

His submarine was similar to the one proposed by William Bourne. It too had a wooden frame which was covered with oiled or greased leather — so it was basically a wooden rowboat wrapped in leather. His 12-man submarine was powered underwater by oars, which were sealed with leather to stop water coming in. This submarine submerged by letting in water, and rose to the surface by pumping it out. He is supposed to have taken King James I for an underwater journey along the Thames, from Westminster to Greenwich, at a depth of 4 to 5 metres. Bourne built three different submarines.

In the next 150 years, there was so much interest in submarines that, in England alone, 14 submarine designs had been patented. But the next documented working submarine was built in one of the British colonies.

The Turtle of 1776

In 1776, during the American War of Independence, an American Yale University

SECRET GAS

One awesome thing about Drebbel's 17th-century submarines is how long they stayed underwater. One submarine was supposed to have stayed underwater for 15 hours! Such a long journey would surely have exhausted the oxygen in the air in the submarine. Some reports talk about a 'secret gas'. Apparently, when the air began to get foul, Drebbel released a gas from leather skins inside the submarine. Of course, this gas would have to be oxygen.

The only problem is that oxygen was not officially discovered for another 150 years! However, Drebbel knew of the work of the Polish alchemist Michael Sendivogius.

Sendivogius had performed many experiments with potassium nitrate, also called nitre. Nitre was easily collected from the walls of sewage pits or latrines. It was used to make gunpowder, so many people were interested in it. Sendivogius had experimented with nitre. He knew that different degrees of heat would make the nitre give off different gases.

We now know that strong heating makes nitre give off fumes of nitrogen dioxide and oxygen, while gentle heating drives off only oxygen. Sendivogius knew of a gas, which he called 'the aerial food of life'. This gas had to be oxygen. Perhaps Drebbel also knew of oxygen, and kept it a secret.

SNORKEL

During World War II, the German submarines introduced the *schnorchel*, which the US Navy called the 'snorkel'. It let the submarine use the diesel engines while submerged. The snorkel was a pipe that rose from the submarine and reached to the surface. It let air for the engines in, and diesel exhaust fumes out.

The snorkel was actually invented by a Dutch naval officer, Lieutenant Jan J. Wichers, in 1933. The Dutch Navy began using them in 1936. Germany invaded the Netherlands in 1940 and began using snorkels. Snorkels enabled the submarines to cruise at periscope depth and so remain hidden, while still being able to run the diesel engines and recharge the batteries. The small radar reflection of the snorkel made them harder to find.

The snorkel was certainly effective. In 1950, a diesel submarine used a snorkel to cruise underwater some 8,370 kilometres, from Hong Kong to Honolulu, in 21 days.

student, David Bushnell, built a tiny egg-shaped wooden submarine (2.3 metres from top to bottom, and 1.8 metres across). There were two hand-cranked propellers — one for forward motion, the other for vertical motion. Lead ballast hanging under the bottom of the *Turtle* kept it upright. It submerged when enough water had been let in via a simple valve, and it could remain underwater for about 30 minutes. This vessel, which Bushnell called the *Turtle*, began the long history of military involvement in submarines. The job of the *Turtle* was to blow up the English flagship, *Eagle*, in New York harbour. The *Turtle*'s one-man crew, Sergeant Ezra Lee, navigated and provided the power.

The plan was that Sergeant Lee would manoeuvre under the *Eagle*, drill into the warship's copper bottom with a long screw, attach a bomb, and then sail away. The bomb had a simple timer to make it explode. The attack happened on the night of 6 September 1776. The screw was easily able to penetrate the copper, but unfortunately for the American rebels, the submarine ended up near the rudder of the British ship. Lee was apparently unlucky enough to hit an iron bracket that supported the rudder. The screw could not penetrate iron, so the attack was not successful.

A replica of Bushnell's *Turtle* was built during the War of 1812 between the United States and the United Kingdom. This time, the pilot managed to screw a hole into the copper sheathing of the HMS *Ramillies*, while it sat near New London, Connecticut. However, the screw broke loose in the process of fastening the bomb to the hull — so once again, the submarine had not yet succeeded as a military weapon.

In 1977, a pair of American historians built a replica of the *Turtle*, and found that with a lot of hard work, a single pilot could make it navigate underwater. The depth at which the *Turtle* cruised was set by the amount of water that the pilot let into the bottom of the submarine. They found that it was very difficult to keep their replica at a preset depth, and that the vertical propeller was virtually useless.

The Nautilus of 1797

In 1797, another American, Robert Fulton, moved to France. He was given a grant from Napoleon Bonaparte to build a submarine. At this time, the British fleet was blockading the French coastline. Fulton was so keen to build his submarine that he actually built it before the French Naval Minister, Forfait, had given him the funds. His submarine, which he called the *Nautilus*, was finished in May 1801. It was about 7 metres long and had sheets of copper laid over iron ribs. Its crew of three or four men propelled the submarine by cranking the propeller by hand.

The *Nautilus* had a few features of modern submarines. It didn't have a full conning tower or sail, but it did have a glass and iron 'bubble' for observations. It had movable horizontal surfaces (what are now called diving planes) to help control the angle at which the submarine cruised. It submerged by flooding a few big ballast tanks with water, but it also had smaller trim ballast tanks (which could also be filled with water) to help control the cruising depth. It even had a streamlined shape to reduce water drag. It was big enough to carry sufficient air to keep the crew alive, and two candles burning, for three hours. Later, he added tanks of compressed air for the crew, and then it could stay underwater for six hours. For travelling on the surface, it carried a folding mast and sail.

Nobody had yet built a self-propelled torpedo. So the *Nautilus* was supposed to sink the ships of the enemy in the same way that the *Turtle* did. In a test, Fulton actually did succeed in sinking an obsolete schooner off the coast at Brest. But when he tried to use the *Nautilus* in battle, the human crew could not get close enough to the British warships. (Some reports say that the *Nautilus* was too slow to catch the British warships. Other reports say that the British had guarded their warships with a circulating fleet of rowboats, under which the *Nautilus* could not penetrate unseen.)

France was no longer interested in Fulton's submarine. So, in the best tradition of arms merchants, he immediately tried to sell his invention to the opposition team, the British. In 1805, in a demonstration to the British, he destroyed the brig *Dorothy* by attaching a mine, which he then blew up with an electric cable. Fulton had the support of the Prime Minister, William Pitt. But Sir John Jervis, the First Lord of the Admiralty, saw Fulton's submarine as a potential threat to the English Navy. The fact that Fulton's submarine actually blew up a ship actually increased Jervis's opposition to the idea of submarines!

Fulton then returned to the USA, and managed to get Congress to back a steam-powered submarine that would carry a crew of 100. He started work on the *Mute*, but died before it was finished. Nobody else could carry through with his vision, and the partly built submarine finally sank at its moorings.

Pop up at your own funeral service

In 1850, the Bavarian inventor Sebastian Wilhelm Valentin Bauer constructed his first submarine, *Le Plongeur-Marin* ('The Marine Diver'). During a test in Kiel Harbour, on 1 February 1851, *Le Plongeur-Marin* sank in 15 metres of water. Bauer

SUBMARINE SUPERLATIVES

Submarines reached their full potential with the introduction of nuclear power. A nuclear power plant meant they could stay underwater for months at a time, while running at full power. According to the 1993 *Guinness Book of Records*, the longest run by a submarine while being continuously submerged was 111 days, from 25 November 1982 to 15 March 1983. During this run, the submarine HMS *Warspite* covered some 57,000 kilometres underwater in the South Atlantic Ocean.

The fastest production submarine is the Russian Alfa class attack submarine. They have been reported to have a maximum speed of 45 knots (83 kilometres per hour), and a maximum diving depth of about 850 metres. The maximum diving depth of most modern attack submarines is between 300 and 450 metres.

The largest submarine in the world is the Russian Typhoon class. They displace some 28,000 tonnes, and are some 170 metres long. These missile submarines carry 20 SS-NX–20 warheads, each of which carries seven independently aimable nuclear warheads.

The largest American submarine is the Ohio class missile submarine. They displace some 19,000 tonnes, and are about 170 metres long. They can carry up to 24 missiles.

and his two colleagues were trapped inside.

The pressure inside the submarine was pretty close to normal atmospheric pressure, while the pressure outside was 1.5 atmospheres higher (the pressure in water increases by 1 atmosphere for each 10 metres of depth). As 1.5 atmospheres works out to 15 tonnes on each square metre, it was impossible to open the hatch. Bauer's crew began to panic, but he did not.

As water gradually leaked into his submarine, the pressure inside slowly increased (yes, their ears would have popped). Bauer hoped that the air would remain breathable until the pressure inside the submarine rose to be equal to the outside pressure. Finally, after a nerve-racking wait of seven and a half hours, the pressure inside was high enough to equal the pressure outside, and the hatch could be opened. The three men easily swam to the surface, and (according to the popular story) popped out of the water in the middle of their own funeral service!

Bauer was not easily frightened, and found another sponsor, the Grand Duke Constantine of Russia. In 1855, Bauer built a 5.8 metre submarine made from iron, *Le Diable-Marin* ('The Marine Devil'). Again, this submarine was human-powered. He took what were probably the first underwater photos through a glass window in *Le Diable-Marin*. He realised some of the future problems that had to be solved, and invented ways of cleaning the air and communicating to other submarines underwater.

To celebrate the coronation of Tsar Alexander II on 6 September 1856, Bauer took some musicians underwater in his submarine in Kronshtadt harbour. People on the surface could hear the musicians playing the Russian national anthem, while his crew sang along!

Unfortunately, this neat trick was not enough. The Russian Navy was very conservative and looked upon the whole idea of a submarine as 'deceitful'. Nobody would support his submarine, so he left Russia in 1858.

Effective weapons of war

During the American Civil War (1861–65), the Northern states blockaded the ports of the South. So the Southern Navy developed small semi-submersibles they called 'Davids' (to fight the 'Goliath' fleet of the North). These semi-submersibles always kept their air intakes and smokestacks above the water line, so they were not true submarines. However, because they were so low, they were hard to see.

The Southern Confederacy also developed CSS *Hunley*, the first submarine to be an effective weapon of war. It had a rocky life.

It was constructed in Mobile, Alabama, from a modified iron boiler, and was carried by rail to Charleston in South Carolina, in 1863. On its first attempted use as a warship, it tried to sink the Northern frigate *New Ironsides*. The *Hunley* was on the surface with the hatch open, and the wash from a passing paddle-steamer sank it — Lieutenant John Payne, the captain, was the only one to escape. The *Hunley* was raised, and in October that year, again tried to sink the *New Ironsides* near Fort Sumter. Again it capsized, but this time the survival rate was better — Payne *and* another crew member! Again it was raised. On 15 October, on a practice dive, it sank again. This time there were no survivors. Horace L. Hunley himself, the Confederate backer of this submarine, died in one of these sinkings.

The *Hunley* was finally 'successfully' used in battle on the night of 17 February 1864, when it sank the Northern corvette,

AROUND THE WORLD, UNDERWATER

The *Triton* was the first nuclear submarine with two nuclear reactors. In 1960, it travelled completely around the world underwater, covering some 67,000 kilometres in 84 days.

The *Triton* followed quite closely the path taken, during the very first circumnavigation of the Earth, by the ship of the Portuguese sailor Ferdinand Magellan. That journey, in the early 1500s, took three years, not 84 days!

Actually, the *Triton*'s trip was not all taken fully submerged. It never rode fully on top of the waves, but it did partly breach the waves twice. The first time was to transfer a member of the crew who was sick, and the second time was at Cadiz, to honour the memory of Magellan.

the USS *Housatonic*, which was blockading Charleston Harbour. Its weapon was a charge of explosive on the end of a 6-metre pole. The *Housatonic* was completely destroyed when its gunpowder magazine also exploded. Unfortunately, the hatch of the *Hunley* was left open, and the resulting wave from the massive explosion sank it, killing all the crew. The *Hunley* was recently discovered, a few kilometres off Sullivans Island.

Engine power

The next big improvement in submarines was an engine. The submarine could never be an effective weapon if it relied on human power.

In 1863, Charles Brun, a French inventor, built a 50-metre submarine, *Le Plongeur*, which used compressed air as a power supply to run an 80-horsepower motor. However, this submarine could not fully submerge (it could only half-dive), and it was also quite unstable. Not only did the compressed air propel his submarine, but it also emptied the water from its ballast tanks, allowing the submarine to rise. This method of emptying ballast tanks is still used today. However, *Le Plongeur* could not store enough compressed air to make it useful as a military weapon.

In 1870, Jules Verne wrote the classic *Twenty Thousand Leagues Under the Sea*, in which Captain Nemo commanded a submarine called the *Nautilus*.

In 1880, George W. Garrett, an English clergyman, ran a steam-powered submarine, the 33-tonne *Resurgam* ('I shall rise'). Obviously, the coal fire that made the steam could not operate while the submarine was submerged. But he could store enough steam under pressure to run the submarine several kilometres underwater. The *Resurgam* would probably qualify as the world's first mechanically powered submarine. After two months of successful trials, it sank while being towed from Birkenhead to Portsmouth, where it was to be further tested by the Royal Navy. It was found in shallow waters, off the coast of Wales, in 1995.

In 1886, two Englishmen, Andrew Campbell and James Ash, built the second (real, not fictional) submarine to carry the name of *Nautilus*. This electrically powered submarine had two 50-horsepower electric motors that could push it at 6 knots (11 kilometres per hour) on the surface. But it had a short range before the batteries ran flat.

In 1888, in France, Gustave Zédé launched another all-electric submarine called the *Gymnote*. This 2-tonne submarine was 17 metres long. It could cruise at 7 knots (13 kilometres per hour) on the surface, and 4.3 knots (8 kilometres per hour) underwater. Like the second *Nautilus*, it too had only a very short range.

In 1899, Maxime Laubeuf, a French Navy marine engineer, launched the *Narval*. This submarine had two hulls, with the ballast tanks in between. This configuration is still used in today's submarines. This 34-metre submarine had two power supplies — steam on the surface, and electric motors underwater.

But not only were combustion engines dangerous in a confined space, the submarine had to stay on the surface practically all the time, to provide air for its engines. This negated most of the stealth philosophy behind the submarine!

The first practical submarine

It was John Philip Holland who developed the first really practical submarine. He built his first submarine in 1875, and spent a quarter of a century refining the concept, until he finally got it right. His company, the J. P. Holland Torpedo Boat Company, was commissioned by the US Navy in 1895 to build a submarine. (In 1900, the name of his company was changed to the Electric Boat Company. Today, it still makes submarines, as the Electric Boat Division of the General Dynamics Corporation.)

He improved his design many times. At his own expense, he built his sixth submarine. This was the one he was satisfied with. It had dual power supplies — batteries for an electric motor for underwater travel, and a petrol engine to run it at seven knots (13 kilometres per hour) while it was on the surface. The petrol engine could also recharge the batteries. The weaponry of this nine-man submarine consisted of a single torpedo tube in the bow (which could fire three torpedoes), and a deck gun that could fire a 45-kilogram shell a distance of about 0.8 kilometre.

In 1900, he delivered his 16.2-metre-long, 70-tonne submarine to the US Navy. The US Navy called it the USS *Holland*. This was the first submarine that could travel underwater for useful distances. Not only did the US Navy order another six of his submarines, and Britain five, but Russia and Japan also placed orders.

J. P. Holland had originally been funded in his submarine building by the Irish Fenian Society of New York, in order to build a submarine that could be used against the British! But, like most weapons makers, he would sell his weapons to anybody. (Actually, the Fenians did not really have their act together. At one stage, one of his submarines was stolen by a rival group of Fenians!)

He used most of the design features that were later used in the submarines of the two world wars. His USS *Holland*

UNPREDICTABLE SCIENCE OF THE SEA

One very important thing about submarine warfare is that our science is not yet knowledgeable enough about the sea.

It sometimes happens that a warship can find, with its sound detection apparatus, a submarine some 50 kilometres away, and not find a submarine only one kilometre away! That undetected second submarine could easily sink the surface warship.

Explosions sometimes behave in the same way. During World War II, a British submarine was towing a few British midget submarines across the North Sea. The midget submarines each carried two side-cargoes of explosives — one on the port side, the other on the starboard side. The plan was that when they were close enough to their target, the German battleship *Tirpitz*, the virtually undetectable midget submarines would attack the *Tirpitz* with their cargoes of explosives.

One of the midget submarines was the X–8. A problem happened, and the starboard side-cargo of explosive was released at sea. It exploded within a few minutes, while it was still within a few hundred metres of the X–8. Surprisingly, there was absolutely no damage to either the X–8 or the submarine doing the towing, so they continued.

Later, the port side-cargo of explosive also had to be prematurely released from the X–8. This time, for safety reasons, a timer was set to two hours. Two hours later, the side-cargo exploded at a range of about 5 kilometres — and the X–8 suffered damage so severe that she had to be abandoned!

submerged by flooding water into its ballast tanks, and rose by blowing the water out with compressed air. To control the depth and cruising angle of the submarine while underwater, he used trim tanks and diving planes. He also used torpedoes, which were fired out of torpedo tubes.

Improvements

The next real improvement was the diesel engine. Petrol fumes were a real explosive hazard in the confined spaces of a submarine, while diesel fumes were not. The first diesel motor to be installed in a submarine was put into a French vessel, the *Aigrette*, in 1905. The German Navy began installing diesel engines in submarines in 1906, and, by 1911, all German submarines had diesel power.

The Germans also introduced high-quality periscopes. Today's modern periscopes are about 12 metres long, but only about 2.5 centimetres across at the top. This gives them a small radar reflection.

World War I

World War I began in August 1914. At that time, the British Royal Navy had the biggest submarine fleet — about 74 submarines. Most of the navies involved in World War I had submarines. Almost all of the submarines were small and slow, and not

THE SPRATLYS

The Spratlys are a motley collection of around 190 sandbars, rocks, cays, atolls, barren islets and islands that are scattered over some 390,000 square kilometres of the South China Sea.

They are not valuable because of what they are — some of them are no more than a largish rock breaking the surface. They are valuable because whoever owns them will also own a 200-mile economic zone around them. This economic zone contains vast wealth, in terms of oil, natural gas and fishing rights. Also, major international shipping lanes pass through this area.

Six countries have laid claim to the Spratlys. China has laid claim to all of the Spratlys. Vietnam, China, the Philippines, Brunei and Malaysia claim parts of the archipelago. China has been systematically building structures on what used to be bare sandbanks mostly submerged by the tides. Their structures have grown into four-storey permanent towers, some of which are equipped with helipads, anti-aircraft guns, air-raid shelters and even plant nurseries.

Vietnam now occupies 25 locations, the Philippines and China each occupy eight, while Malaysia occupies just one islet. Brunei has not grabbed any land.

very useful militarily, so they were given only coastal duties. The exception was the German Navy.

Early on, the German naval command realised that submarines could be used most effectively as a strategic weapon, rather than a tactical (battlefield) weapon. German submarines sank some 12 million tonnes of shipping, and almost cut the United Kingdom's trade links across the Atlantic.

The German U-boat (*Unterseebooten*, or 'boat that travelled under the sea') was relatively sophisticated. In one single incident on 22 September 1914, the submarine U–9, under the command of Kapitän-Leutnant Wedigen of the Imperial German Navy, sank three British cruisers in less than 45 minutes. Some 36,000 tonnes of British warships were lost, as well as some 1,200 men. Everybody suddenly realised that the submarine was an incredibly effective threat to surface ships.

The British retaliated by trying to keep the German submarines in their ports (by laying mines), and by hunting the U-boats at sea. They did this by developing underwater listening devices to find the U-boats, and depth charges to destroy them. They also developed the very effective R class of anti-submarine attack submarines that could travel at 15 knots (28 kilometres per hour) underwater for two hours! The British also gathered the merchant ships into groups that were well guarded by military escort vessels. This convoy tactic proved effective until the end of World War I.

Between the wars

Between the world wars, there was not much progress in submarine technology, apart from improved reliability and living conditions. Many remarkably silly ideas were tried, and abandoned.

Both the French and the British built submarines in which large deck guns were the main armament. The enormous French submarine *Surcouf* displaced 4,373 tonnes and carried two 8-inch guns.

The French, Japanese and British navies experimented with submarines that carried seaplanes. The British M.2 sank in 1932, when the hangar door for the seaplane was left open.

World War II

During World War II, submarines were again incredibly effective. Although they were not much faster than the submarines of World War I, they sank almost as much tonnage of aircraft carriers as were sunk by aircraft. Altogether, some 20 million tonnes of ships were sunk.

Again, German submarines in the Atlantic Ocean tried to stop maritime supplies to the United Kingdom, and again, they almost succeeded. They sank some 4,770 ships. The small VII-type U-boat was the main German submarine used in the Atlantic. Typically, they displaced about 760 tonnes, were about 67 metres long, and could travel at 17 knots (31 kilometres per hour) on the surface, and 7.5 knots (14 kilometres per hour) underwater.

The British responded to the submarine threat to their transatlantic trade with extra air patrols to attack the U-boats, and by

SUPER TORPEDO

Russia has developed a torpedo that will travel underwater at the awesome speed of some 200 knots (370 kilometres per hour)! This is about 140 knots (260 kilometres per hour) faster than previous torpedoes. The Russians call this torpedo the *Shkval* (or 'Squall').

The traditional formula for the maximum speed that a vessel can travel in water, if it has access to an infinite power supply, is this: the speed (in knots) = 1.34 multiplied by (the square root of the length of the waterline, measured in feet).

As you can see, the longer the waterline, the faster the boat. To travel at 200 knots, a torpedo would need, according to this formula, a waterline some 9 kilometres long!

As this is currently impossible, the clever Russians have come up with another approach. Their torpedo does not travel in water. It travels under the surface of the water, but it's surrounded by a continuous wall of a gas, which is generated by a small engine inside the torpedo! Gases have much lower friction than liquids, and so the torpedo can travel at this incredible speed. The Russians are currently modifying it to travel at 300 knots (555 kilometres per hour).

There will be lots of countries that would just love to buy this torpedo for their submarines.

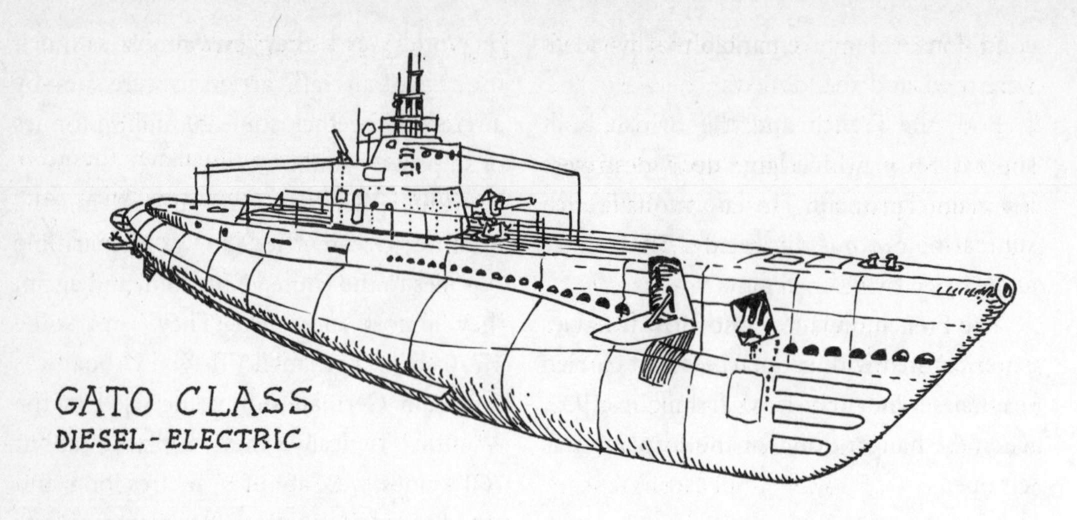

using the tactic of the heavily guarded convoy. The German Navy retaliated by forming 'wolf-packs' of many U-boats, which were co-ordinated by long-range aircraft (to find the convoys) and radio. The Allies responded with more air patrols and aircraft carriers, and more attack submarines to chase the U-boats. It was very close, but the U-boats did not succeed in cutting the trade across the Atlantic Ocean.

However, the American submarines in the Pacific Ocean did succeed in destroying the Japanese merchant navy. By mid–1944, they had effectively isolated Japan. The two main submarines used by the US Navy were the Gato-class (which could dive to about 90 metres) and Balao-class submarines (which could dive to about 120 metres). They displaced about 1,500 tonnes, were about 95 metres long, and could maintain 20 knots (37 kilometres per hour) on the surface, and about 9 knots (16 kilometres per hour) underwater. (It's strange, but the Gato-class submarine was longer than the depth that it could dive to!)

British and American submarines in the Mediterranean Sea successfully stopped the German Navy from delivering supplies to the German troops in North Africa.

There was not much difference in performance between the various submarines used in World War II. The average US submarine could travel on the surface at about 18 knots (with diesel engines) and at about 8 knots underwater (with electric motors).

Fish-shaped submarines

One long-term problem with the standard long, skinny shape of most submarines was that at certain high underwater speeds, the submarine might suddenly pitch up or down, and control would be lost.

It took a surprisingly long time to realise that perhaps Mother Nature had got it right the first time, and that the shape of the fish was a good design to copy.

The next big step came when the Americans discovered, with their blimp-shaped *Albacore* submarine in 1952, that the long, skinny cigar shape was not the best shape for a submarine.

This was actually a gamble that paid off handsomely. The Cold War between the USA and the USSR was getting quite tense. The US Navy had to give up a destroyer conversion program, so as to have the money to build an experimental submarine.

This submarine had no weaponry at all. It was built purely to test the different shape. It turned out that a fairly short, blunt front end, and a longer, tapering rear end, had much less drag. Suddenly, with the same size of power supply, a submarine could travel at 30 knots (56 kilometres per hour) underwater, not 20 knots (37 kilometres per hour). This fish-like submarine could also be turned much faster — about 3.2 degrees per second, as compared to less than 2.5 degrees per second for the old, skinny cigar shape. For the first time, a submarine had so much manoeuvrability that it could actually be 'flown' underwater. It's even rumoured that it once did a complete 360 degree loop! This new shape also meant that there was more useful storage space.

Nuclear power

The next big step was nuclear power.

The first nuclear submarine, the USS *Nautilus* (the third submarine with that name), was launched on 17 January 1955. It then sent the famous signal: 'Under way on nuclear power.' The *Nautilus* was some 98 metres long, displaced some 3,230 tonnes, carried a crew of 105, and could cruise at over 20 knots (37 kilometres per hour) underwater. In one of its first trials, it sailed underwater some 2,170 kilometres from New London, Connecticut, to San Juan, Puerto Rico, in just 84 hours. It had the same cigar-like shape as conventional submarines, but its great range transformed it into a remarkable weapon of war. While its first nuclear power pack took it some 100,000 kilometres, the third nuclear power pack propelled it some 240,000 kilometres. On 3 August 1958, under Commander William R. Anderson, the *Nautilus* sailed under the North Pole.

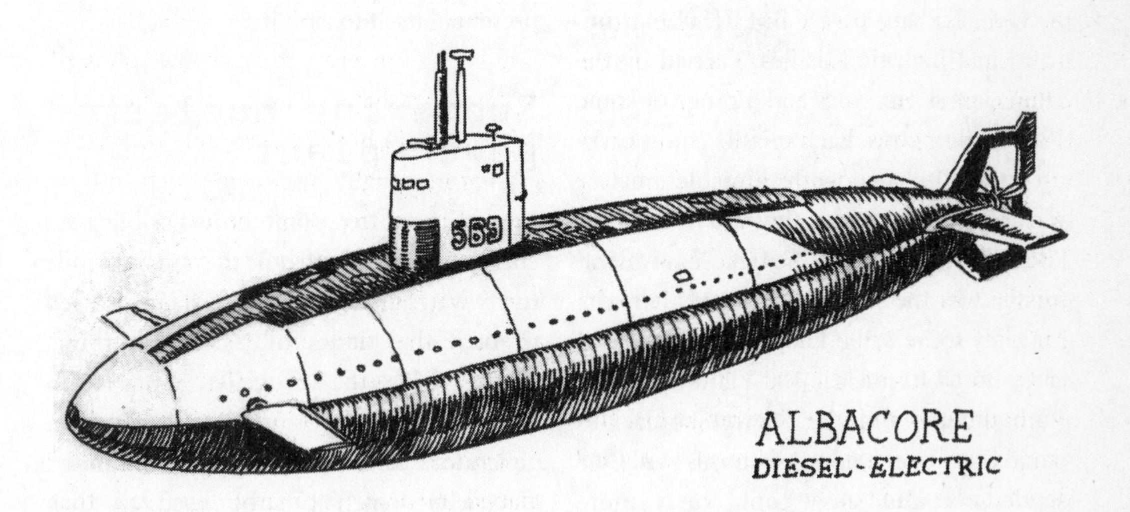

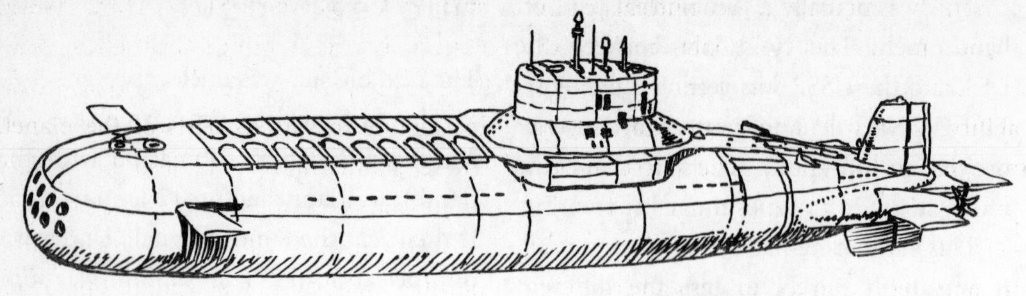

Soon after, the US Navy combined the *Albacore*-shape with nuclear power to develop the Skipjack-class of nuclear submarines. The USS *Skipjack*, launched in 1959, could travel underwater at over 30 knots (56 kilometres per hour)!

There are two main types of nuclear submarines: missile submarines and attack submarines.

In 1960, the USS *George Washington* launched a Polaris ballistic missile while it was underwater. This led to the most powerful single weapon of destruction ever invented by the human race — the nuclear-powered, ballistic missile submarine.

The Trident II SLBM (Submarine-Launched Ballistic Missiles) carried by the Ohio class submarines had a range of some 12,000 kilometres. Each missile could carry up to 12 independently aimable nuclear warheads, each containing up to a 150-kiloton bomb. The Soviet 'equivalent' missile was the SS-NX–20 SLBM. It had a range of some 8,300 kilometres, and could carry up to 10 nuclear 100-kiloton bombs. Both the US and the Soviet submarines could carry over a dozen such missiles. One single submarine now could carry more explosive power than was used in all of World War II. One single nuclear-powered, ballistic missile submarine had the potential to destroy up to 10 per cent of the human race!

The other type of nuclear submarine is the attack submarine.

In the Falklands War in 1982, the Argentinians pulled all of their ships out of the area after a British nuclear submarine, the HMS *Conqueror,* sank the fast Argentine cruiser *General Belgrano.* This was a superb example of the clever use of a nuclear submarine. Only a nuclear submarine would have the underwater capability to follow a fast surface ship for 48 hours, before the right opportunity presented itself to sink it.

Submarine nuclear power plant

A nuclear power plant is just a big heat engine. The heat from the nuclear pile turns water into steam. The steam is used to turn the blades of a steam turbine, which turns the propeller. The steam turbine also powers an electrical generator. This gives virtually unlimited electricity which can be used to turn seawater into drinking water, and to clean up the air.

The advantage of the nuclear power plant is that it suddenly makes the submarine a true underwater ship, with the ability to travel underwater for several months at full speed. The latest nuclear power plants can run a submarine for over 600,000 kilometres. The time it can stay underwater is limited by human frailty — the crew needing new supplies.

The disadvantage of the nuclear power plant is that it's noisier than a diesel-electric submarine. Pumps have to run all the time, to keep various working liquids moving. You can't just switch the nuclear reactor off when you want to make your submarine quiet. The best you can do is throttle the reactor right back down to idle, but still the liquids have to circulate, and this makes some noise.

When a diesel-electric submarine is floating quietly underwater, there are no moving parts — apart from the spinning hard drives in the computers. (If you have enough RAM in the computers, you don't even need hard drives!) The electricity to power the submarine comes from batteries, which are also silent. So the loudest noise in the diesel-electric submarine would be the noise of people.

Cold Wars

During the recently finished Cold War between the USA and the then-USSR, the nuclear-powered submarines at sea that were carrying the nuclear-tipped ballistic missiles were the real deterrent to all-out nuclear war. Once these submarines slipped out to sea with their deadly load and submerged, they were virtually undetectable.

That Cold War is over. Now we are building up the weapons for a new set of smaller Cold Wars. Many Third World countries are arming themselves with submarines, which could be a great threat to stability in various parts of the planet. These submarines could also be a very significant threat to some very large navies.

Today, nuclear submarines are still the ultimate political status symbol. But diesel submarines have one great advantage — they can be very quiet.

According to a military analyst, 'a submarine is a hidden force that produces maximum effect for minimum effort.'

In a recent joint naval exercise between Australia and the USA, one of Australia's diesel Oberon class submarines was able to sneak up and smack a dummy torpedo into the belly of an American aircraft carrier.

Let's consider the possible use of diesel submarines in one recent war. Think about the narrow mouth of the Persian Gulf, the Strait of Hormuz. About one-quarter of the world's oil passes through that narrow choke point. In the recent Gulf War, Allied naval vessels moved freely through the Persian Gulf. But according to US Vice Admiral James Williams, if Iraq had bought six modern diesel attack submarines and 'positioned three of them on either side of the Strait of Hormuz, that would have complicated matters ... One diesel sub can make a great difference to how you drive your ships.'

In the spring of 1993, Iran began to show off the first of its Russian-made Kilo-class diesel attack submarines in the Persian Gulf. The USA quickly responded by putting two Los Angeles-class nuclear-powered attack submarines in the Gulf.

Think back to the Falkland Islands War. One single Argentine German-supplied Type 209 diesel submarine, the *San Luis*, created havoc in the British fleet. Even though the British knew she was somewhere nearby, they could not stop her. The *San Luis* managed to avoid 15 British destroyers and frigates that were chasing her. She also eluded the combined anti-submarine aircraft of two separate aircraft carriers. The *San Luis* managed to get within effective torpedo range of the British fleet, to fire three torpedoes. But all three shots were unsuccessful!

The Falkland Islands War was a close thing. Imagine how differently that war might have turned out if those three torpedoes had hit.

Back in the Cold War, nuclear-tipped ballistic missiles were a real worry, because they could easily penetrate enemy defences in a surprise attack. Attack submarines also easily penetrate enemy defences in a surprise attack. This is what makes them so attractive. Unfortunately, arms manufacturers see submarine sales as a market opportunity.

Subs and weapons for sale

At the moment, over 20 Third World countries are operating some 150 diesel attack submarines. And they all want more. They range from North Korea which has 25 diesel attack submarines, down to Libya and Pakistan, which each have six. Most of them are not effective fighting vessels — being obsolete, badly maintained, or run by badly trained crews. But you need only one effective diesel attack submarine to make a world of difference.

Germany has been making Type 209 diesel attack submarines for about 20 years — but not one of them was sold to the German Navy! They have all been exported. Germany has also set up 'co-production' agreements with Argentina, India and South Korea. This means that these countries actually make some components of the submarines. Argentina has a 'licence to re-export'. This means that Argentina can make more Type 209 submarines and sell them. Ultimately, such a licence means that today's customer can turn into tomorrow's competitor for sales.

The USA has recently set up an unusual agreement involving Type 209 diesel attack submarines. The USA stopped making diesel subs in the 1960s. Egypt wants two more Type 209s, but doesn't have the money to buy them. However, in April 1994, the US State Department gave permission to Ingalls shipyard in Pascagoula, Mississippi, to make two Type 209s. Egypt will 'buy' these attack subs, using US military aid (which can be spent only on American-made weapons).

The other odd thing about this agreement is that the US Navy doesn't want it. The US Navy wrote, in a 1992 Report to Congress: 'Construction of diesel submarines for export in US shipyards would not support the US submarine ship-building base and could encourage future development and operation of diesel submarines to the detriment of our own forces.'

Taiwan and Saudi Arabia could well be next to buy Type 209s off the American production line.

One big problem with selling weapons is that they can be used against the seller. When the Shah of Persia was overthrown, his country (Persia, now Iran) had six Type 209 submarines on order. If Iran had had these submarines, how different the power balance in the Persian Gulf would be today.

PIG-BOATS

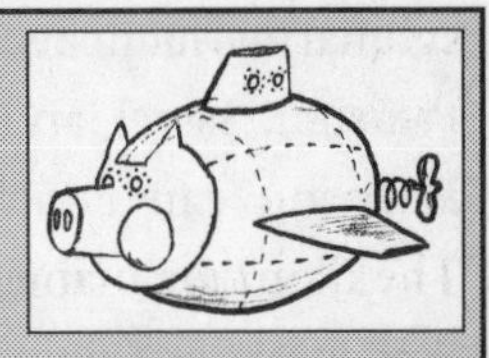

Diesel submarines have long been called 'pig-boats' by the men who sail in them. This is because they smell so bad!

Power and clean water are very limited in a diesel submarine when it's cruising underwater. So once a diesel sub slips out to sea on a patrol, nobody does a full body wash. They all begin to smell bad at the same rate, so nobody complains. In World War II, some Japanese submariners would actually scrape dead skin off each other with the sharp edges of wooden boards.

Nuclear-powered submarines smell much better. They have all the power and clean water anybody could possibly use.

Russia has just sold eight Kilo-class submarines to the Chinese. This will add to its fleet of 45 diesel attack submarines. These submarines could easily put a blockade around Taiwan. Taiwan already has four diesel attack submarines, and wants 10 more. Taiwan has approached Australia to buy some submarines. A deal for 10 submarines would be worth about $6.5 billion. However, the Clinton administration in Washington did not want to compromise its relationships with China by selling submarines to Taiwan. Australia, as a traditional military partner to the USA, will most probably not sell submarines to Taiwan.

South Korea already has six Type 209 submarines, and wants to buy another four of them — at a cost of US$1 billion.

Japan already has 16 submarines, and plans to build another 12.

France has sold Daphne-class and Agosta-class submarines to Pakistan. The Agosta-class submarines would cost around US$233 million each.

China has sold some of its older Romeo-class submarines to Egypt and North Korea.

Sweden is selling submarines to Malaysia. However, Sweden has guidelines which prevent it from selling weapons to countries which it believes are violating human rights. So Sweden, which designed the Australian-made Collins-class submarines, may try to block the sales of these submarines to Indonesia. The Collins-class submarines displace about 3,000 tonnes, and are about 78 metres long. It will cost about $4.75 billion to build just six of them.

Indonesia already has two submarines, and has ordered three more from Germany.

Meanwhile Britain is so keen to pension off four of its Upholder-class submarines (displacing about 2,000 tonnes, and about 70 metres long), that it will even lease them, complete with crews, to whoever wants them.

Of course, the problem is not just the diesel attack submarines, but the weapons that they are fitted with.

Already, the wealthy countries have supplied these poor countries with super-advanced torpedoes that will explode immediately under the target ship, so as to cause the maximum amount of damage. The USA has already sold the Harpoon submarine-launched anti-ship missile to Pakistan, Israel and other countries. The Harpoon can attack a ship as much as 145 kilometres from the launch site. The French are trying to sell their marine versions of their Exocet missiles. We all know how effective Exocets were in the Falkland Islands War.

Currently, diesel submarines can stay under water for only a short time — a week or so. But during that week, they can be very hard to find in the shallow waters near a coast. Not only are there 'false sonar echoes' from the bumpy sea floor, there are sunken ships, also called 'magnetic garbage', to confuse the magnetic sensors of the hunters.

And shipyards in South Korea, Russia, Germany, Italy and Sweden are already working on propulsion systems which would let a submarine stay underwater for a month.

Today, submarines are the biggest potential threat to commercial shipping. In a future war involving submarines, they would use guerrilla battle tactics. They would remain hidden until they suddenly struck an unprepared enemy. Submarines would strike in the enemy's weak spots, and then would suddenly disappear back into the ocean.

This recent, and enormous, proliferation of diesel attack submarines could lead to a series of nasty Cold Wars between squabbling neighbours. Unfortunately, selling submarines is good money — the Chinese are spending some US$2 billion on their eight kilo-class submarines. And unfortunately, many of the sales contracts involve the licensing of the designs, and the buyers getting the technology to build more submarines.

Toward the end of the Cold War, it was decided by the USA and the USSR that it was just too dangerous to have so many nuclear weapons on the planet. So they began to disarm. But now we have a new weapons race in diesel attack submarines.

One foolish aspect of this problem of the proliferating submarines is that many countries buying submarines are horrendously poor, and have massive health problems. According to a recent United Nations report, they could solve all their health problems with just 11 per cent of their total military budget! But they insist on buying weapons, instead of vaccines.

Now, selling submarines is good money. But perhaps there is an argument for balancing short-term profits of the arms makers against long-term political stability.

REFERENCES

Emsley, John, 'A gas by any other name', *New Scientist*, no. 1966, 25 February 1995, pp. 38, 39.

Encyclopaedia Britannica, 1996.

'Fear over Russian rocket torpedo', *Sydney Morning Herald*, 25 May 1995, p. 9.

Leary, Joseph, 'The Turtle dives again', *Invention & Technology*, Spring 1966, pp. 18–26.

Revelle, Daniel J. & Lumpe, Lora, 'Third World submarines', *Scientific American*, August 1994, pp. 16–21.

DAYLIGHT SAVING CAUSES ACCIDENTS

Yet another way daylight saving causes accidents.

Australia was one of the first countries in the world to adopt Daylight Saving. Once a year, as we head into summer, most Australians put the clock forward by one hour. A few months later, as we leave summer, we wind the clocks backward.

But now there's a report that claims that the change-over to Daylight Saving leads to a temporary increase in road accidents! And it's probably because most of us are permanently sleep-deprived!

Hours of sunlight each day

The number of hours of daylight (or sunshine) depends on two things: the time of the year, and where you are on the planet.

During the summer and winter Equinoxes (usually around 21 March and 23 September), everybody on the planet gets 12 hours of sunlight and 12 hours of night. (*Equi* means 'equal', and *nox* means 'night', so the Equinox is when the number of hours of darkness, or night, is the same as the number of hours of sunlight.)

At the Equator, there are 12 hours of sunlight all year long. But at the North Pole, around Christmas, you get 24 hours of darkness and zero hours of sunshine. At the same time, at the South Pole, you get 24 hours of sunlight and zero hours of darkness, which is called 'The Midnight Sun'.

And at locations between one of the Poles and the Equator, the situation is somewhere in between, depending on the time of year. For example, Sydney is about 34 degrees of latitude from the Equator. At the height of summer, around 23 December, there are 14.5 hours of daylight, and only 9.5 hours of darkness.

Daylight Saving isn't for everybody

Daylight Saving Time 'works' when you get lots of hours of sunlight. Its aim is to take one of those hours of sunlight away from the morning, and to relocate it at the end of the day.

Under Daylight Saving Time, the clocks are put forward, usually by an hour, in summer — so that 7 pm is suddenly called 8 pm. Suppose that on one particular day, at 7 pm, there are 90 minutes of sunlight left. So sunset *should* happen at 8:30 pm. But when you switch to Daylight Saving Time, the sunset gets 'shifted' one hour later, to 9:30 pm.

So, while the total number of hours of sunlight stays unchanged, Daylight Saving Time gives an extra leisure hour of daylight to the 9-to-5 workers, at the end of the working day. They can do with it what they will.

But Daylight Saving Time is really effective only at the higher latitudes. It works better the further you are away from the Equator. After all, it would be useless at the Equator, where you get 12 hours of sunlight all year round.

War brings Daylight Saving

Daylight Saving was first suggested, almost light-heartedly, by that great All-American All-Rounder, Benjamin Franklin, back in 1784. But at the time, nobody really took him seriously.

In 1907, an English builder, William Willett, wrote an essay called 'Waste of Daylight'. He proposed setting the clock forward by 80 minutes, in four separate jumps of 20 minutes each, as the calendar moved through spring into summer. The next year, in 1908, a bill was put to the House of Commons to advance the clock by one hour in springtime, and to wind it back to Greenwich Mean Time in the autumn. The bill was rejected.

The desire for extra leisure time wasn't enough to get the world to try Daylight Saving Time. But the pressures of war were much more persuasive.

Daylight Saving Time was first adopted in World War I to save fuel, by cutting down on night-time artificial lighting. Several countries, including the United States, Great Britain, Germany and Australia, adopted Daylight Saving. But in Australia, it proved so unpopular after it was introduced in 1917, that it was abandoned the very same year.

During World War II, Daylight Saving Time was introduced into the USA and the UK on a full-time basis, where it was called 'War Time'. The US had continuous Daylight Saving Time from 9 February 1942 to 30 September 1945. Not only did England have continuous Daylight Saving Time for the duration of the war, but during the summer months it had 'Double Summer Time', where the clocks were advanced by another extra hour.

Why we (don't) need Daylight Saving

There have always been many arguments for, and against, Daylight Saving Time.

The main advantage was the increased leisure and family time at the end of the working day. Entrepreneurs suggested that sales of sporting equipment (necessary to enjoy this leisure time) would probably increase as well — which would be good for business. Other claimed advantages included savings in electrical power and, it was hoped, a reduction in road accidents. This would supposedly happen because there would be fewer people on the road during the night, when accidents are more likely to happen.

Disadvantages of Daylight Saving Time included the claim that it lengthened the working day of country people by one hour. It would also supposedly upset the daily routine of farm animals (such as laying hens and cows) that follow the Sun rather than the clock. Parents of young children would presumably have difficulty in putting their children to bed. (As a parent of young children, I would agree. On the other hand, sometimes, it's not that easy to imagine that job could be any harder!)

But one angle to the Daylight Saving Time discussion is something hardly anybody thought of — sleep.

Over the last 10 years, our working hours have increased by 10 per cent. Many of us are permanently sleep-deprived, and never wake up feeling rested.

We already know that insufficient sleep, and disrupted circadian rhythms (which shift workers on rotating rosters experience all the time), can cause major health problems. They also cost a lot of money. In 1988, the cost of accidents related to sleep problems amounted to more than US$56 billion.

Some major accidents which have been linked to insufficient sleep and/or disrupted circadian rhythms, include the destruction of the space shuttle *Challenger*, the nuclear accident at Chernobyl, and the oil spilled from the ship *Exxon Valdez.*

In 1988, as a result of sleep-related accidents, about 25,000 people died, and 2.5 million people suffered disabling injuries.

Currently, about 25 countries shift to Daylight Saving Time every spring, and the next autumn return to their standard time. Each time they shift into Daylight Saving Time, everybody loses one hour of sleep time. And when they shift out of Daylight

DAYLIGHT SAVING TIME IN AUSTRALIA

In Australia, the Federal Government introduced one-hour Daylight Saving Time into all Australian states in 1942. Daylight Saving remained in force until 1944, although Western Australia had already pulled out earlier, on the grounds that it caused inconvenience to workers who followed the Sun rather than the clock.

That was the end of Daylight Saving in Australia until 1967, when Tasmania introduced one-hour Daylight Saving Time. Tasmania, the furthest state from the Equator at a latitude of 44 degrees, would benefit more from Daylight Saving Time than the mainland states.

In 1971, after much controversy and discussion, one-hour Daylight Saving Time was introduced into New South Wales, Victoria and the Australian Capital Territory. Queensland, much of which is in the tropics and which would therefore not get as much benefit from Daylight Saving, reluctantly joined in, but soon abandoned Daylight Saving. In 1974, Western Australia adopted one-hour Daylight Saving Time, but it was thrown out by a referendum in 1975. New South Wales decided to adopt Daylight Saving Time after a referendum in 1976. Even though many of the rural areas were strongly opposed to Daylight Saving, they were out-voted by the more populous city areas.

Saving Time into standard time, everybody gets one extra hour of sleep time.

Micro-sleeps on the road

Stanley Coren, from the University of British Columbia, decided to see whether the change-over, into and out of Daylight Saving Time, had any effect on traffic accidents. He and his team had access to the data on the 1,398,784 accidents that were reported to the Canadian Ministry of Transport during the years 1991 and 1992. They looked at the accidents that happened on the Monday before the changeover, and the two following Mondays.

They found that when Canada went into Daylight Saving Time in the springtime (and when people lost one hour's sleep), there was an 8 per cent increased risk of accidents on the Monday immediately after the changeover. One week later, on the next Monday, the risk of accidents had dropped back to almost the normal rate, and was only about 1–2 per cent higher than usual.

But when people had one hour's extra sleep (when they shifted out of Daylight Saving Time back into normal time), there was a 6 per cent reduced risk of traffic accidents on the Monday immediately after the change. And a week later, there were about 1–2 per cent more accidents — again, virtually the same as being back to the normal accident rate.

So on average, shifting into Daylight Saving increases the accident rate (for a little while) by about 8 per cent, while shifting out

DAYLIGHT SAVING TIME OVERSEAS

In 1966, the US passed the Uniform Time Act. Within each time zone of the US, this Act defined Uniform Daylight Saving Time. Of course, in the Land of The Free, not all of the states followed this recommendation. At first, Daylight Saving Time lasted from the last Sunday in April to the last Sunday in October. The duration of Daylight Saving Time was temporarily extended during 1974 and 1975, thanks to the famous Gasoline Energy Crisis. In 1986, legislation was passed to change this period so it would begin at 2 am on the *first* Sunday in April and cease at 2 am on the last Sunday in October.

Most of the countries of Western Europe have a slightly different schedule — from the last Sunday in March to the last Sunday in September. Britain is different again — from 30 March to 26 October.

of Daylight Saving gives us a roughly similar 6 per cent reduction in accidents.

One explanation for this is that we are all so sleep-deprived that anything that takes away an hour of sleep makes us less competent. Coren thinks that the shift into Daylight Saving Time, 'might lead to an increased number of "micro-sleeps", or lapses of attention, during daily activities, and thus might cause an increase in the probability of accidents, especially in traffic.' The accident rate is almost back to average by the second Monday after the change. However, according to Coren, 'Measurable changes in sleep pattern persist for up to five days after each time shift.'

Unfortunately, Coren's study is just one single study looking at just three weekends in two years, in one country. It would be good to see some follow-up studies on this topic. First, can other workers in the field actually duplicate these results (in other words, to see if this is a real effect, or just a statistical accident)? Second, perhaps the study could be broadened, for example by looking at other countries to see how long the 'change-over' effect lasts for, and so on.

But, according to this one study, getting one hour's less sleep can increase your chances of a car accident the next day by 8 per cent.

Apart from all the arguments about Daylight Saving Time, maybe the real message is that we all need more sleep.

REFERENCES

'Adjusting to the changes to, and from, Daylight Saving Time', *Science*, vol. 261, 1976, pp. 688–689.

The Australian Encyclopaedia, Australian Geographic (for the Australian Geographic Society), Sydney, 1988, p. 964.

Coren, Stanley, 'Daylight Saving Time and traffic accidents', *New England Journal of Medicine*, 4 April 1996, p. 924.

'The cost of sleep-related accidents: a report for the National Commission of Sleep Disorders Research', *Sleep*, vol. 17, 1994, pp. 64–93.

Encyclopaedia Britannica, 1996.

My nipples have gone all jealous.

MEN'S NIPPLES & BREAST-FEEDING

Over the last 20 years or so, there's been a big change in our society. Men are getting more involved in childcare. But while men can take the baby for a walk, and change the baby's nappy, there's one thing that they definitely can't do — breast-feed. Well, sometimes they can — and this might answer the ancient question of why men have nipples!

Goats and fruit bats

Now, it's not impossible that a male animal can grow mammary glands, and then give milk, or lactate. For example, every now and then, you'll come across a male goat that can lactate. There's a herd of goats in Bathurst, west of Sydney, where the male goats are regularly milked by the goatherds. These are perfectly normal goats, with normal male genitals, who have fathered baby goats, and yet they can breast-feed, or udder-feed, their young. But in general, male mammals can't usually feed their babies with milk.

There are about 4,500 species of mammals. Until recently, the zoologists had not found a single species where the males *regularly* have breasts, and give breast milk to the babies. But this all changed in February 1994. That's when Doctors Francis, Anthony, Brunton and Kunz from Boston University published a report about the Dayak Fruit Bats (*Dyacopterus spadiceus*) of Malaysia. Not all the male fruit bats that they captured were lactating, but neither were all the females. It depended a lot on the season. If there were baby fruit bats waiting to be fed, many of the adult male bats had mammary glands that were swollen with breast milk. The male bats also had perfectly normal testicles that were making normal sperm.

So now we know that there exists at least one (and so far, only one) species of mammal, where the males regularly feed their young with breast milk.

NIPPLES

In mammals, both the males and the females have mammary glands. Mammary glands are always more developed in the females — but there is a range of difference. For example, mammary ducts and nipples don't form in mice and rats. But in humans, other two-legged animals and dogs, mammary ducts and nipples do form. In human and other two-legged animals, before puberty, there is no real difference between the species in the mammary tissue.

The female breast is divided into 15 to 20 lobes. Blood comes into these lobes, and breast milk comes out. Each lobe has a tiny duct (or pipe) for the milk to come through. These 15 to 20 ducts all lead to the nipple, where they appear as tiny holes (about 0.3 millimetres in diameter) arranged in a circle.

But if men don't generally breastfeed, then why do they have nipples?

Men and women are incredibly similar, and have the same body plan, and there are only very small differences between them. Their chromosomes (where the DNA is) are the same, except for the two sex chromosomes. Women have XX for their sex chromosomes, while men have XY.

Some people say that since a Y is just an X with one of the arms knocked off, men are just the economy model, while women are the fully optioned human! The embryologists recognise that as the fertilised egg goes down the pathway that leads to a human baby nine months later, the female body is the 'basic' human body, and the male body is the 'modified, fewer options' human body.

So one reason that males have nipples is that the making of nipples is 'wired' into the basic software of the DNA, and that it would cause too much upset to the system to delete the nipples in males. Another reason is nipples are an erogenous zone in both men and women — and it's only fair for both sexes to have a good time!

Extraordinary conditions

Under normal conditions, human males don't lactate. There are a few, very rare, diseases that can make men lactate, and men can sometimes lactate if they're taking hormones. Men with some cancers are often treated with the female sex hormone, oestrogen. In the past, some of these men were also given another hormone, prolactin — and many of them produced milk from their nipples!

Starvation can also make men lactate. During World War II, many millions of men were starved in Nazi and Japanese concentration camps. At the end of the war, the Allies swept through and released the prisoners. Allied doctors noticed that several thousand of these starving men had grown breasts, and that these breasts were delivering milk. In one single Japanese concentration

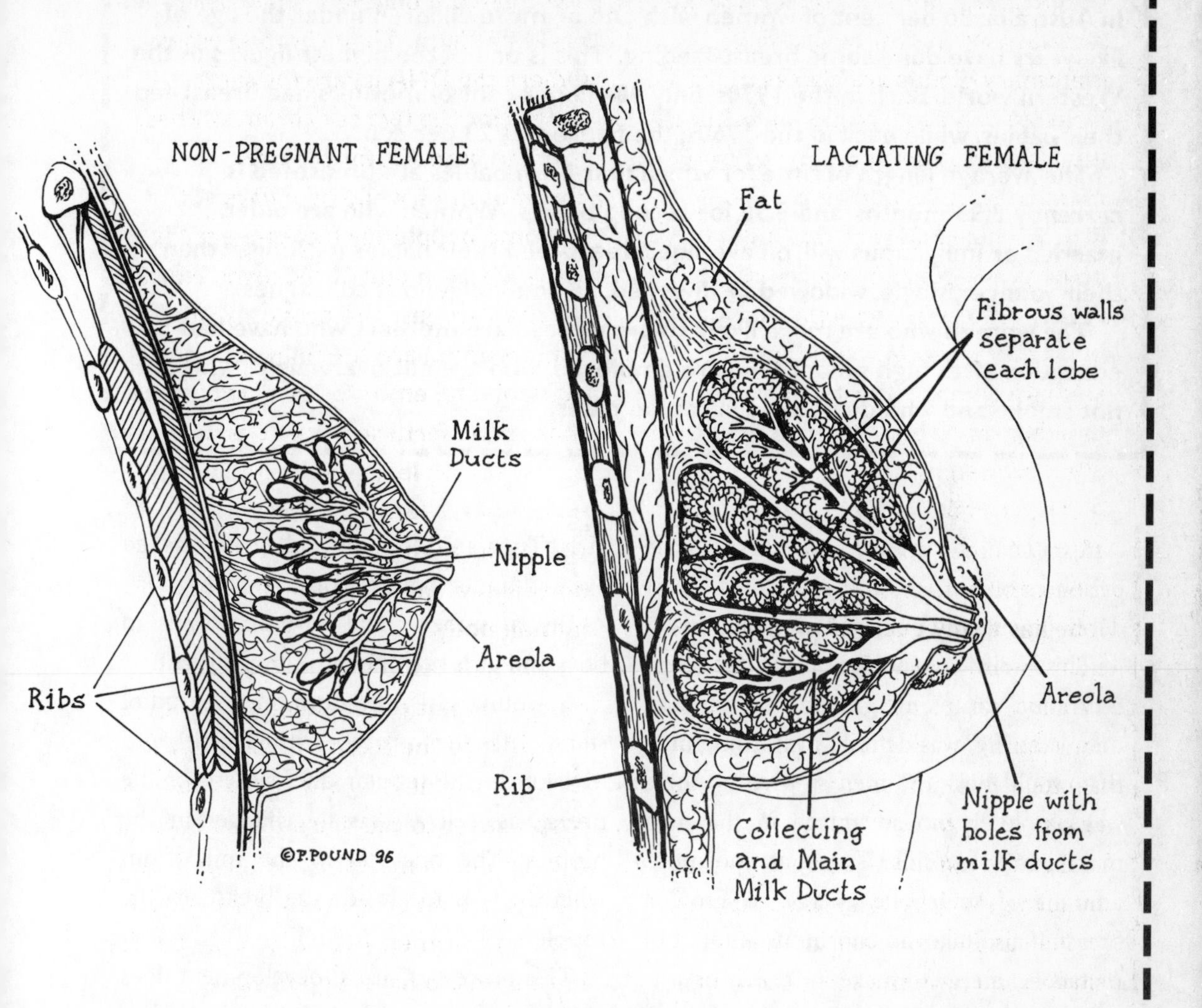

WHAT IS A MAMMAL?

The breast is basically a modified sweat gland that turns blood into milk. It's also one of the body parts that says humans are one of the 4,500 species of animals that are called mammals.

Other things that define us to be mammals include the following: we're warm blooded, we bear living young that feed on mother's milk, we have large brains, we have at least some hair somewhere on our body, we have sweat glands in our skin, we have ears on the outside of our bodies, each side of the lower jaw is made up of a single bone, we have a heart with four chambers and, finally, in our ears we have three middle-ear bones.

WHO BREAST-FEEDS?

In Australia, 80 per cent of women with one or more children under the age of five years have done some breast-feeding. This is one of the highest figures in the Western world. Back in the 1970s, only 40–45 per cent of mothers had breast-fed their babies, while back in the 1960s, the figure was 23 per cent.

The average length of time for which first-born babies are breast-fed is currently 8.53 months, and 8.99 for second babies. Women who are older, married or indigenous will, on average, breast-feed their babies for longer than their younger, single, widowed or divorced, or non-indigenous colleagues.

The women who are more likely to breast-feed are mothers who have had education after high school, who are employed, who are not overweight, who do not smoke and who do not live in capital cities.

camp alone, there were 500 starved male prisoners-of-war who had significant breast development, and who were lactating.

This weird phenomenon in the concentration camps, of growing breasts and then lactating, was definitely *not* something that would help the men survive — there was absolutely no advantage in the men making milk to drink. It takes an enormous amount of metabolic energy to grow a breast. It also takes an enormous amount of metabolic energy to make milk. Any breast-feeding mother can tell you that breast-feeding a baby really strips the fat off you. So growing breasts, and then making milk, would actually have brought the starving men a lot closer to death. The men in the concentration camps lactated because of a hormonal imbalance induced by starvation.

Hormones

Male breast growth, and lactation, is all to do with hormones.

Before puberty, there is no observable difference between the breasts of boys and girls. They are identical both to the naked eye and under a microscope.

But at puberty, the hormone levels in boys and girls become radically different.

In young girls, the hormone oestradiol starts off the process of growth and division in the tubular duct system of the breast, as well as causing changes in the nipples. The ducts carry the milk from where it is made inside the breast, to the nipple.

This process of breast development does not happen in young boys, because they don't make oestradiol. In fact, even if you were to give boys oestradiol, they would not grow breasts, because they have other hormones floating in their bloodstream.

But you can start breast development in boys by giving them other hormones. The breasts that they grow are exactly the same as girls' breasts. These hormones are the female sex hormones, oestrogen and progesterone, and they cause breast development when they are given in a ratio of somewhere between 1:20 and 1:100.

So you can 'grow' a breast in a human

male, providing you give him the 'right' hormones. But to make milk, you need a bunch of other hormones. The most important of these is prolactin ('lactin' as in 'lactose' and 'lactate').

Women will normally lactate when there's simultaneously a high level of the hormone prolactin, and a low level of the hormone oestrogen. This usually happens immediately after her baby is delivered.

CONSIDER THE ECHIDNA

Injections of various female sex hormones have grown mammary glands and produced milk in male guinea pigs, male goats and a steer! If you give ovarian hormones to male echidnas, their breasts will grow larger than the breasts of female echidnas, and will then deliver lots of milk.

Wet nurses and 'keen' babies

Breast-milk production can also happen in other circumstances. The phenomenon of the 'wet nurse' is well known. This is when a non-pregnant woman will be able to breast-feed a baby, after a few weeks of nipple stimulation by the baby's sucking. It turns out that stimulation of the nipple directly causes an increase in prolactin levels in the blood.

Suppose a woman adopts a baby. Suppose that she places the baby on her breast, and lets it suck. If she does this regularly for about a month, she will begin to deliver breast milk to her new baby. The 'wet nurse' phenomenon can happen in grandmothers some 71 years old, and even in women who have never been pregnant. It's even mentioned in the Bible, where Naomi, Ruth's mother-in-law, was able to act as a wet nurse.

Virginia Thorley Phillips, a Brisbane lactation consultant, has been involved in a few unusual cases of 'keen' babies. These babies had been weaned from the breast, for periods up to 6 months. These clever babies had somehow convinced their mothers to let them have a little suck, and

NOT A QUESTION OF SIZE

Women with smaller breasts can breast-feed perfectly well. There are two major components inside the breast: tissue that makes milk, and fat. So the breast-feeding capacity is not always obvious, just by looking. Women with smaller milk capacity will just breast-feed their baby more frequently.

According to some studies, breast-feeding does not change the shape of the breasts — time does. Women who have breast-fed babies have the same range of size and shape of breasts as women who have been pregnant but who have never breast-fed. So, according to some studies, pregnancy does change breast shape, but breast-feeding doesn't.

THE BENEFITS OF BREAST MILK

Breast milk is wonderful stuff. The average breast-feeding woman can generate about a litre of breast milk each day. This litre contains a lot of water, 70 grams of lactose, 38 grams of fat and 12 grams of protein. Various chemicals in breast milk, such as fat, calcium and other important nutrients, are better absorbed by the baby than from infant formula. Breast milk also contains antibodies from the mother, as well as various important hormones and growth factors that formula milk doesn't have.

Breast milk also has DHA, one of the long-chain omega–3 fatty acids (very popular in the current press). DHA helps in the building and operation of brain cells, and is also taken up by cells in the retina. This could be why babies fed on breast milk have, in the first eight to nine months of life, sharper vision than babies fed on infant formula. DHA, and the other long-chain omega–3 fatty acids, can't be added to infant formula, because they oxidise very rapidly to make nasty chemicals.

Lactoferrin is another chemical in breast milk. Lactoferrin is a type of iron that is very well absorbed by a baby. It also may have anti-infection properties.

shortly afterwards, milk began flowing from the breast again!

Sometimes, stimulation of the male nipple is enough to make milk, if the stimulation is frequent enough. Dr Basil Donovan, Director of the Sydney Sexually Transmitted Diseases Centre at Sydney Hospital, has seen this a few times. The male breasts exude colostrum, a milk very rich in protein.

Perhaps this is why a tribe in Brazil was reported to have men who breast-fed. M. Howard, in *Victorian Grotesque*, writes about this. According to missionaries of the 16th century, there was a tribe in Brazil in which the women had 'small and withered breasts and whose children were brought up from birth by suckling the males'.

Little glands and littler liver

So how did the starved men in the concentration camps grow breasts and lactate?

First, you need to know that men and women make the same hormones, but in different amounts. Consider the female sex hormone, oestrogen, and the male sex hormone, testosterone. Women make lots of oestrogen and hardly any testosterone, while men make lots of testosterone and hardly any oestrogen. But the essential thing is that they do make the same hormones.

Second, the level of any hormone in your blood depends on two factors — how much is being made, and how much is being broken down. In the body, various hormones are constantly being made by various glands, and these hormones are always being broken down and recycled — usually by the liver.

GARLIC-FLAVOURED BREAST MILK

In the old days, breast-feeding mothers were advised not to eat food with strong flavours — foods such as garlic, onions and cabbage. But today, the advice might be completely different.

Professor David Laing (Professor of Food Technology at the University of Western Sydney) and his colleague Champa Jinadasa studied a group of 30 babies aged between four and six months. The babies were in three sub-groups: breast-fed by women who ate lots of garlic, breast-fed by women who ate only a little garlic, or fed infant formula from a bottle. The test food for the babies was pumpkin flavoured with garlic.

The results were as expected. The formula-fed babies ate the least garlic–pumpkin mix; whereas the breast-fed babies with garlic-eating mothers ate the most of the garlic–pumpkin mix, were quicker to start eating the mix and, by the end of the study, were eating twice as much of the mix.

So if, as your kids grow up, you want them to eat what you do (so you don't have to cook extra meals), guzzle on the spicy foods while breast-feeding!

The level of a hormone in your blood is a bit like the level of water in a sink.

Suppose you have a tap which lets water into a sink, and a drain hole which lets the water out of this sink. The level of water in the sink depends on how much water is flowing into the sink, and how much is flowing out.

Suppose that the tap is just a little bit open, so only a small amount of water is flowing into the sink. But suppose that the drain hole is almost totally blocked, so that very little water is flowing out. The water level will slowly rise in the sink. As it gets higher, there will be greater water pressure at the bottom of the sink, which will increase the rate at which the water flows out. Eventually the water level will stay steady at a fairly high level.

The same kind of situation was happening to the starved men in the concentration camps. Their glands had shrunk a bit, so they were making fewer hormones than normal (like our tap being 'on' only a little bit). But in these starving men, the liver had shrunk enormously. The tiny liver was hardly able to break down even the small amounts of various hormones that were being made (like the blocked drain hole that would let only a very tiny amount of water out). So the level of some hormones in their bloodstreams rose a long way above the normal levels.

So, in some of these millions of starving men in concentration camps around the world, the hormones that rose to abnormally high levels were the hormones needed to grow breasts, and to make milk.

When these lactating men were given normal diets, the milk production actually increased for a while. That was because their glands recovered very rapidly and made even more hormones. But the liver took much longer to recover, so

IS BREAST-FEEDING A GOOD CONTRACEPTIVE?

The elderly female domestic engineer's story (old wives' tale) is true — breast-feeding can be a good contraceptive.

The World Health Organisation (WHO) undertook an enormous study that involved some 8,000 women, from 15 countries, covering the years from 1989 to 1993. WHO found that if women fully breast-feed for six months after their baby was born, then their chance of becoming pregnant during that time is less than 2 per cent!

This is an amazingly effective contraceptive method — about as good as the Pill, and much better than a condom (which gives a woman up to 15 per cent chance of getting pregnant).

However, to achieve 98 per cent reliability as a contraceptive, three conditions have to be met. First, the woman must have no periods. Second, this 98 per cent reliability applies only for the first six months after the baby is born. Third, her baby must be getting all (or practically all) the nutrition needed from her breast milk. If the baby is getting other sources of food, this means that she is not delivering enough breast milk to achieve the 98 per cent reliability.

for a little while, there was nothing breaking down these hormones. Their hormone levels briefly went through the roof!

Male childcare

There are about 4,500 species of mammals in the world today. Of these species, 90 per cent have very uncaring fathers — fathers who don't even hang around to gather food, or teach their growing babies tricks, or even defend the territory or their family.

But 10 per cent of the species of mammals on our planet have lots of male parental involvement — species such as lions, gibbons and even humans. In these 10 per cent of species, it is occasionally possible that male lactation occurs.

The mammals have been dominant on our planet only since the death of the dinosaurs, some 65 million years ago. Could it be that, long ago in our evolutionary past (say, 50 million years ago), both male and female mammals fed their young from their mammary glands? Could it be that we have only recently (say, 30 million years ago) become specialised, so that today it's usually the female mammals who feed the young?

Maybe we humans will one day follow the example of the Dayak Fruit Bat of Malaysia. Maybe in a more politically correct, caring-and-sharing world, that has a well-developed genetic engineering technology, the blokes (just like the goats at Bathurst) will be able to feed the baby breast milk.

And maybe that explains the mystery of why males have nipples (apart from pleasure) — those useless buttons, which

EMBARRASSED BY BREAST-FEEDING?

According to Dr Martha Morrow, a scientist at the Melbourne University Key Centre for Women's Health, some women don't want to breast-feed because of some of the attitudes of some parts of society.

In a paper presented to the 14th Annual Australian Perinatal Conference in Adelaide, in March 1996, she said that many women stop breast-feeding when their baby is about six weeks old, because they are 'uncomfortable breast-feeding around others, or find it distasteful. Women's embarrassment may be related to advertising and pornography's use of the breast to display eroticism. It is no wonder some women feel awkward about having their baby suckle on a body part which is a central focus of pornographic magazines and films.'

Dr Morrow went on to say: 'Not all societies find breast-feeding embarrassing. In Indonesia — a Muslim country with strict standards of female modesty — women commonly suckle infants in public, and make no attempt to be discreet or cover themselves. Yet in Australia where we pride ourselves on being an open society it is common to hear of women being asked to stop breast-feeding in public.'

In the very litigious USA, the State of Florida actually had to legislate on breast-feeding. It has recently passed laws saying that breast-feeding does not constitute sexual conduct, is not harmful to minors, and does not violate prohibitions.

have long been a source of embarrassment to teenage males — they were just waiting for men to jump into the nitty-gritty of suckling children!

REFERENCES

Archer, Michael, 'Coming to grips with male nipples', *Australian Natural History*, vol. 23, no. 6, Spring 1990, pp. 494–495.

Diamond, Jared, 'Father's milk', *Discover*, February 1995, pp. 82–87.

Francis, Charles M., *et al.*, 'Lactation in male fruit bats', *Nature*, 24 February 1994, pp. 691–692.

Howard, M., *Victorian Grotesque*, Jupiter Books, London, 1977.

Huggins, Charles & Dao, Thomas L.-Y., 'Lactation induced by luteotrophin in women with mammary cancer. Growth of the breast of the human male following estrogenic treatment', *Cancer Research*, vol. 14, 1954, pp. 303–306.

Juan, Stephen, *The Odd Body*, Angus & Robertson, Sydney, 1995, pp. 128–130.

Morrow, Martha, paper presented to the 14th Annual Australian Perinatal Conference, Adelaide, March 1966.

COMING!
Ready or not...

OTHER EARTHS, OTHER SOLAR SYSTEMS

Astronomers are hot on the trail of other solar systems. If they can find other solar systems, they're just one small step closer to finding intelligent life out there.

It's been a long search, but by the end of 1995, astronomers were pretty sure that they had found a handful of planets fairly similar to Jupiter, circling a handful of stars very similar to our own Sun. This is very exciting stuff!

Studying the Universe

Thinkers have long wondered about our place in the Universe.

Three centuries before Christ was born, the Greek philosopher Epicurus proposed that in our Universe, there should be an infinite number of planets. He predicted that some of them would be like our Earth, while others would be quite different.

Around the same time, the Greek astronomer Aristarchus of Samos claimed that the Earth was part of a solar system. As part of this concept, he also claimed, rather radically, that this system had the Sun in the middle. He was correct, but it was not until the 16th century AD that

HOW SINGLE STARS FORM

The most popular theory today on the origin of the solar system is the so-called 'nebular hypothesis'. It says that the solar system came into existence, about 4.5 billion years ago, from a single large rotating cloud, or nebula. Amazingly, this theory was put forward around 200 years ago, by two people working quite independently of each other. In 1755, the German philosopher Immanuel Kant discussed the origin of the solar system in his book *Universal Natural History and Theory of the Heavens*. In 1796, the French mathematician Pierre Laplace also put forward a very similar theory in his book *Treatise on the World System*.

Now, the space between the stars is not completely empty. Floating in space are giant clouds of various atoms and molecules. Astronomers have even found giant clouds of vinegar, and alcohol! Some of these clouds are fairly thin, while others are very dense.

The nebular theory says that for some reason or other, parts of the cloud begin to contract. For example, a shock wave from a nearby exploding star, or supernova, would be enough to set off a contraction in a cloud. Once one part of the cloud begins to contract, it becomes denser and more massive, which gives it a greater gravitational field — so it will tend to 'suck' more matter onto itself.

Eventually, this giant spherical cloud of gas and dust begins to rotate. As it rotates, the spherical cloud tends to flatten out into a spinning disc. At the very centre, a 'blob' of matter begins to accumulate rapidly. It begins to heat up — both from gravitational collapse, and from the kinetic energy of falling gas, dust and rocks.

If the 'blob' at the centre weighs less than 8 per cent of the mass of our Sun, it never quite gets hot enough to 'burn' hydrogen as a nuclear fuel. But if the blob has more than 8 per cent of the mass of our Sun, it will 'burn' hydrogen, and the star will soon flash into nuclear fusion.

philosophers and astronomers accepted his proposal.

Four hundred years after Aristarchus, the ideas of another Greek, Ptolemy, *were* accepted by educated people. Unfortunately, his ideas were wrong! In the early 2nd century AD, he wrote that the Earth should lie at the very centre of a Universe which is shaped like a ball. Orbiting around the Earth, he said, were the Sun, the Moon, and the five known planets (Mercury, Venus, Mars, Jupiter and Saturn). It took over 1,000 years before another major thinker, the Polish astronomer Nicolaus Copernicus, publicly disagreed. Copernicus suggested the notion of a solar system with a Sun at the centre — and other thinkers began to accept his concept. The Church, however, took a lot longer to see the truth.

Our Solar System is beautifully even and symmetrical

Our Solar System has nine planets: four small rocky planets close to the Sun, four big gas planets further out, and finally, a small odd planet called Pluto.

The four inner planets — the so-called 'terrestrial planets' (Mercury, Venus, Earth and Mars) — are small and are mostly made of rock and metal, although they do have fairly insignificant atmospheres.

The next four planets out from the Sun — the so-called 'gas planets' (Jupiter, Saturn, Uranus and Neptune) — have small-ish cores, but most of the planets' huge bulk is taken up by very dense atmospheres that are fairly rich in hydrogen and helium. The gas planets range from 14 to 318 times heavier than the Earth.

Pluto, the ninth planet, is quite odd. Pluto is usually described as a planet with a moon, but there are actually two planets in close orbit around each other. Pluto is four times heavier than its companion, Charon. This is incredibly close — most planets are hundreds of times heavier than their moons. So Pluto and its 'moon', Charon, should really be thought of as a small double-planet system.

Our Solar System is beautifully even and symmetrical. All the planets move around the Sun in the same direction, and most of them orbit in almost perfectly concentric circles. (However, for part of Pluto's orbit around the Sun, it actually cuts inside the orbit of Neptune.) The orbits of the planets (apart from Pluto) are all on virtually the same plane — fairly closely lined up with the equator of the Sun. The four inner planets

HOW PLANETS FORM

Planets can pop into existence incredibly rapidly.

As the ball of gas and dust rotates, it gradually flattens out to form a disc, like two soup plates stuck together. This can take as little as 10 million years.

The gas and dust drift towards the mid-plane of the disc (halfway between the top and the bottom). If a lump of dust is about 1 micron (one-millionth of a metre) across, it will take an average of about 100,000 years to reach the mid-plane. But if it's 1 centimetre across, it will get there in just 10 years. Once the lumps of dust are all together in the mid-plane, they begin to stick to each other, and grow. They are more likely to stick together if they are moving in the same direction with roughly the same speed. If they run into each other at high velocity, they often bounce off and away from each other.

Computer simulations show that objects as big as 10 kilometres across will then coalesce in as little as 1,000 years. These objects, called 'planetismals', are the basic building blocks of the planets. These planetismals run into each other and stick together, until they form a planet.

We think that the inner planets would have taken about 20 million to 100 million years to 'make', after the ball of gas and dust began to collapse. Then, in our Solar System, the left-over debris bombarded the surfaces of any planets or moons for about 600 million years.

LITTLE CLOSE ROCKY PLANETS VERSUS BIG DISTANT GAS PLANETS

Why are the little rocky planets close to the Sun, while the big gas planets are distant from the Sun?

One theory deals with the pressure of the solar wind. It says that very near to the Sun, there is so much radiation and solar wind thrown out from the Sun that the gas is stripped away from any bodies that do form. This supposedly explains why the inner planets of our solar system are mostly metal and rock, with hardly any gas. But further out in the solar system, the outward pressure of the solar wind is much less, and so gas planets can form.

Another theory deals with the heat of the Sun. It says that the materials that could survive high temperatures (such as rock) condensed and formed near the heat of the Sun. The materials that can survive only at lower temperatures formed further out from the Sun.

And why are the gas planets (mostly hydrogen) so big, while the rocky planets (mostly rock, made from silicon and oxygen) so small? Because in the Universe, there's lots of hydrogen but not much silicon. In fact, the ratios of hydrogen to oxygen to silicon are 24,000 to 1,000 to 1. So because there's only a small amount of silicon to start with, the planets made of silicon are small.

The discovery of our new planets has caused some anxiety about these theories. Pegasus 51 has a Jupiter-sized planet orbiting closer than Mercury. But it shouldn't! Astronomers have suggested that it formed at the 'standard' distance, but that it later migrated in, via gravitation interactions with the disc of dust from which it came. Back to the drawing board ... maybe.

are all small, and mostly made of metal and rock, while the four outer gas planets are all huge, and mostly made of gas.

Our current theories of how our Solar System formed can explain all this. What these theories can't tell us is how to 'find' other solar systems.

We know a few different techniques to find other solar systems: interferometry; looking for pancake-like discs of dust and gas around a star; looking at the 'wobble' in the motion of a star; looking for changes in the timing of radio signals naturally emitted by a star; and looking for a change in the velocity of a star.

Interferometry

Simply using a telescope may seem an obvious way to look for other solar systems, but you immediately come across two big problems: closeness and glare. First, from our great distance away, any supposed planet would appear to be very close to its central star. Second, it would also be much duller — about a billion times duller! This is because planets do not 'make' their own light, but merely reflect the light of a nearby star. Trying to see a planet next to its star is like trying to see the yellowish glow of an almost-flat torch next to a billion-watt searchlight! The 'glare factor' is enormous, and the light reflected from the planet would be swamped by the light emitted by the star.

Under these conditions, to see a planet with a telescope we would need the telescope mirror to be as big as a football field — which would be very big, very fragile and very expensive!

To get around this problem, scientists use a technique called 'interferometry', which mimics the effect of a giant mirror. Luckily, you don't actually need to make the giant mirror.

Imagine that you have a curved mirror as big as a football field. Imagine that you cut out two small discs, each a metre in size, on opposite sides of this mirror. Then throw away the rest of the football-field-sized mirror, while still keeping the two small mirrors in their original positions and alignment.

If you can somehow electronically, and optically, link the light that falls onto these two small mirrors, you end up with a rather special telescope. It has the light gathering power of two one-metre mirrors. But it has the sharpness, or resolving power, of a mirror the size of a football field! This kind of linked telescope system is called an interferometer.

There are a few big interferometers on the way. The 10-metre Keck telescope on Mauna Kea, in Hawaii, is currently the biggest telescope on the planet. When the second 10-metre Keck telescope is fully

HOW DOUBLE STARS FORM

About 70 per cent of the stars in the Milky Way are not single stars, but double stars — two stars going around each other.

The theory to explain how double stars form is slightly different from the theory for single stars.

Once again we have the huge cloud of gas and dust floating free between the stars. Once again, a shock wave comes through and begins to compress parts of the gas cloud. But this time, the cloud breaks into two fragments each of which begins to shrink down, quite independently of each other. If they're massive enough, each part of the cloud will contract to form a separate star — which is how you end up with two stars, or a double star system.

Besides their origins, there's another major difference between our Solar System and double star systems. In our Solar System, the planets go around the central Sun in a very circular orbit. Double stars, on the other hand, tend to go around each other in a very elliptical or egg-shaped orbit.

THE SEARCH INSIDE OUR SOLAR SYSTEM

The astronomers of ancient Egypt knew six of the nine planets in our Solar System: Mercury, Venus, Earth, Mars, Jupiter and Saturn.

The seventh planet out from the Sun, Uranus, was the first planet to be discovered with a telescope. William Herschel found it in 1781.

The eighth planet, Neptune, was discovered as a result of perturbations (or 'bumps') in the orbit of Uranus. Sometimes Uranus would 'speed up' in its orbit, and at other times, it would 'slow down'. Several astronomers and mathematicians were convinced that there was another planet out there, pulling on Uranus. Depending on its location, sometimes this supposed eighth planet would pull Uranus forward, and at other times, pull it back.

The French mathematician Urbain-Jean-Joseph Le Verrier gave his

predictions as to the position of the eighth planet to the astronomer Johann Galle. In 1846, Galle used the telescope at the Berlin Observatory to discover Neptune.

But after a while, the astronomers thought that they had found 'bumps' in the orbit of Neptune. Once again, predictions were made. And so, in 1930, Clyde Tombaugh discovered the ninth planet, Pluto — and immediately afterwards, to celebrate, he went out to see a movie called *The Virginian*.

However, it was all a colossal coincidence! Pluto turned out to be much too small to cause the supposed perturbations in the orbit of Neptune. In fact, according to recent studies, there never were any 'bumps' in Neptune's orbit! Tombaugh was just plain lucky.

linked to the first 10-metre Keck telescope, some 85 metres away, the combination will be the most powerful interferometer on the planet. It will have the light gathering power of two 10-metre mirrors, but the resolution of an 85-metre mirror.

NASA has a much more ambitious design on the way. It plans, for a few hundred million dollars, to put a kilometre-long interferometer in orbit by the year 2020. It is hoped that 'The Planet Finder' will see lots of detail. NASA expects to be able to pick up not only Earth-sized planets close to a distant star, but even continents and oceans on those planets!

This all sounds wonderful, but as of 1996, nobody has actually seen a planet around a distant star — neither with an ordinary telescope, nor with an interferometer.

Discs of dust and gas

A second technique to find other solar systems again uses telescopes. This time astronomers look for pancake-like discs of dust and gas around a star.

Currently, we believe that a disc is a temporary stage in the formation of a solar system. If we can see such a disc around a star, it very probably means that planets are forming, or will soon form. The Hubble Space Telescope easily has the resolving power to see such discs around young stars.

In June 1994, Robert O'Dell, from Rice University in Houston, Texas, and Zheng Wen, from the University of Kentucky, announced the results of their search with the Hubble Space Telescope. They used the Hubble to look at 110 stars in the Orion Nebula.

The Orion Nebula is about 1,500 light years away. It's a regular little nursery, where baby stars just pop into existence. These stars are condensing out of gas and dust.

The Orion Nebula is incredibly densely packed with stars. For comparison, the nearest star to Earth (other than our Sun) is about four light years away. But in the Orion Nebula, there are 4,000 stars in just one cubic light year!

O'Dell and Wen reported that half of the 110 stars that they observed in the Orion Nebula had discs around them! That's a lot of potential planets.

Other scientists have found discs, but with circular gaps in the discs.

Back in 1992, two astronomers announced something unusual that they had seen when they looked at T Tauri stars. T Tauri stars are young stars that will evolve into stars that are very similar to our own Sun. Kenneth Marsh and Michael J. Mahoney, from the Jet Propulsion Laboratory in Pasadena, California, looked at several T Tauri stars some 450 light years away.

They saw gaps in the discs around eight of these stars. Five of these gaps are quite large, and probably are caused by 'brown dwarfs' sweeping up the gas and dust as they grow. (A brown dwarf is like a 'failed' star, and is bigger than Jupiter but not quite big enough to ignite into nuclear fusion and shine.) But three of the gaps were quite small — too small to be caused by brown dwarfs. They suggested that these gaps could be made by planets that were sweeping up the gas and dust.

In another survey of young stars, 60 per cent had a dust disc around them. Thirty stars had discs which ranged in size from 100 to 1,000 AU. (An AU, or Astronomical Unit, is the distance between the Earth and the Sun.)

The minimum mass needed in a dust disc, to build the planets of a solar system, is about 1 per cent of the mass of our Sun — and most of the dust discs that we have seen are bigger than this. Not all of these discs will make planets, but even so, there should be a lot of planets out there.

Back in 1984, the IRAS infra-red orbiting telescope photographed a disc around the star Beta Pictoris, about 50 light years away. This dust cloud is made from silicates, water, methane, ammonia and iron — the stuff from which our own planets are made. This dust disc also has a gap — which very probably means yet another planet.

In January 1996, Christopher Burrows, from the Space Telescope Centre in Baltimore, announced that he had found that the dust disc of Beta Pictoris was slightly warped. He suggested that a planet, in the gap between the inner edge of the disc and the star, is causing this warping by its gravitational effect.

Dip in brightness

A third technique for finding planets is to watch for a temporary dip in the brightness of a star as a planet goes across the face of the star. On average, about 1 per cent of the possible planetary systems would appear edge-on to us. A planet the size of Earth, going across the front of a star the size of our Sun, would cause a 0.01 per cent dip in brightness. As yet, however, no planets have been discovered by using this technique.

Wobble

A fourth technique used is to look at the 'wobble' in the motion of a star as it moves across the sky.

Most people imagine that the Earth sails majestically in a smoothly curved orbit around the Sun, and that the Moon goes around the Earth. Neither of these 'facts' is true!

Instead, both the Earth and Moon orbit around their common centre of gravity, which is about 1,400 kilometres beneath the equator of the Earth. This common centre of gravity is what goes around the Sun in a smoothly curved orbit. So both the Earth and the Moon rotate around this centre of gravity — sometimes above it, and sometimes below it.

Now an observer on Mars would find it hard to see the Moon, and would see only the Earth. So, over a period of several months, an observer on Mars would see Earth appear to 'wobble' in its orbit around the Sun.

Some of the nearby stars move noticeably across the sky in a few decades. We would expect that the same kind of 'wobble' would happen with one of these stars and its planet(s) — if it had any. However, we would expect the effect to be quite small, because the star is usually very much heavier than its planet(s).

Imagine a bicycle wheel with a weight on one part of the rim. When you spin it, the bicycle wheel wobbles. In the same way, a star with a heavy planet in orbit around it would appear to wobble as the star moved slowly across the sky. This technique works only on stars which are quite close to us, and which move fairly rapidly across the sky.

Barnard's Star is one such star. It's the second-closest star system to us. It was discovered in 1916 by Edward Emerson Barnard. This star is a small red dwarf, only 14 per cent of the mass of the Sun. It's quite close — about six light years away. It appears to move across the sky very rapidly, and will cover a distance equal to the diameter of a full moon in only 180 years. That is a huge distance, when you consider that most of the stars in the sky appear not to move over thousands of years. Astronomers at Sproul Observatory in Philadelphia began looking at Barnard's Star in 1938.

In 1962, Peter van der Kemp announced that Barnard's Star appeared to wobble across the sky. In 1963, he followed up by writing that it probably had a planet in an elliptical orbit around it. He calculated that this planet was about 1.6 times the mass of Jupiter, and took 24 years to do a complete loop of Barnard's Star. In 1969, he updated his findings, and claimed that it had two planets. He said that one planet, at 70 per cent the mass of Jupiter, orbited at 2.7 AU, and took 12 years for a complete loop. The other supposed planet, at 50 per cent the mass of Jupiter, orbited at 3.8 AU, and took 20 years for a complete orbit.

When other astronomers repeated van der Kemp's work, however, both on his and other telescopes, the general consensus was that the observed wobble was not 'real', but related to 'problems' with his telescope.

PREDICTING OUR SOLAR SYSTEM

Many of the fine details of what we know about our Solar System come from the Voyager 1 and Voyager 2 spacecraft. They each sent back close-up photos of Jupiter and Saturn. Voyager 2 continued on to explore Uranus and Neptune. They made astounding discoveries.

This was a chance for scientists to make predictions about the gas planets and their moons before the spacecraft arrived, and then check the predictions against the data sent back. Andrew Prentice, a mathematician at Monash University in Melbourne, Australia, has been right in his predictions more often than anyone else!

He successfully predicted the rocky ring around Jupiter. He successfully predicted that Tethys, a moon of Saturn, would be 22–25 per cent heavier than previously believed. It was 21.5 per cent heavier. He successfully predicted that Voyager 2 would discover two new moons of Uranus. He successfully predicted their locations, chemical compositions and density.

He did all this with his modern version of Laplace's Theory. One specific detail of Laplace's Theory is that the contracting gas cloud would cast off successive rings of gas as it shrank. These rings would eventually condense into planets. This detail implies that the planets did not all form at the same time, but that the outer ones formed before the inner ones. In turn, each gas planet would cast off its own rings, which would, in turn, condense into moons.

Prentice has a variation on Laplace's Theory which he calls 'supersonic turbulence'. Many of Prentice's colleagues won't accept 'supersonic turbulence' — but in a scientific sense, he is the one picking the winners. He has not been proved wrong on any of the gas planets in our Solar System!

A further test of his theory would be to try it out on other solar systems — as we discover them.

A similar 'discovery–undiscovery' situation happened with a star called Van Biesbroeck 8, or VB 8. VB 8 is about 21 light years away from us. Back in 1984, some scientists claimed that they had found a wobble in its movements. In 1987, however, the existence of this wobble was disproved.

As of 1996, no planets had yet been found with the basic wobble technique.

Radio signals

A star and its invisible companion will orbit about their common centre of gravity. So the fifth technique uses this information and looks for changes in timing of the regular radio signals naturally emitted by the star.

Luckily for the astronomers, there are some stars that do emit incredibly regular radio signals. These stars are not your 'normal' star — these stars weigh as much as our Sun, but are only about 15 kilometres in diameter! For comparison, our Sun is about 1.5 million kilometres in diameter.

On their surface, these odd stars have a 'hot spot' that blasts out radio waves. As the star rotates, it emits radio waves in exactly the same way that a lighthouse emits light beams. We can pick up the radio 'blips' only when the 'hot spot' on the star is facing directly at us. These stars can spin at up to 1,000 rotations per second. They are called 'pulsars'. Their names begin with PSR, which stands for Pulsating Source of Radio.

The current theory is that pulsars are what is left behind when a star explodes, or goes 'supernova'. A pulsar is the ember of a supernova.

When a star goes supernova, it blasts out, in one second, as much energy as all the stars (about 400 billion) in a galaxy emit in their entire lives (about 5 billion years)! That's a lot of energy! Astronomers used to feel that any planets around that star would be destroyed, or at the very least, be thrown out of orbit by the colossal explosion. It turns out that the astronomers' feelings were wrong.

Normally, pulsars are incredibly regular in the timing of their radio signals. So any change in the timing of the 'blips' is easy to measure. Light travels at about 300,000 kilometres per second. If the star is one kilometre closer, then the signal arrives one–300,000th of a second early. Suppose that there is a planet around the pulsar — so both the pulsar and the planet rotate around their common centre of gravity. When the star is closest to us, the radio signals arrive a tiny bit early. When the star is furthest away from us, the radio signals arrive a tiny bit late.

If, over a period of time, the astronomer finds regular delays, and then advances, in the timing of the radio signals sent from a pulsar — bingo, a planet!

In July 1991, Matthew Bailes, Andrew Lyne and Setnam Shemar announced that they had found a planet orbiting a pulsar some 30,000 light years away. They claimed that they had found a six-month variation in the arrival time of radio pulses coming from the pulsar known as PSR 1829–10. This pulsar emits three radio signals per second.

Six months is a very suspicious time! It's the time taken for the Earth to travel from closest to the pulsar, to farthest away. The astronomers knew this, and compensated for this factor in their calculations.

Unfortunately, the astronomers had not compensated for the fact that the orbit of our planet, Earth, is not perfectly circular — but is actually elliptical. All they had discovered was our own planet!

Andrew Lyne announced the mistake in January 1992, at a meeting of the American Astronomical Society, in Atlanta, Georgia. His voice was choked with emotion, as he told the audience: 'This talk is not the one I was originally proposing to give ... Our embarrassment is unbounded, and we are sorry.' He received thunderous applause and a standing ovation, for the admission of an honest mistake.

The editor of *Nature*, Mr Maddox, wrote: 'It is to be hoped that they will not be too downcast by this turn of events. On the contrary, they should take pride in the directness of their acknowledgment this week of their mistake ... The world at large should understand that such ups and downs are inseparable from science.'

By an amazing coincidence, at the same meeting, Alexander Wolszczan, from Arecibo Observatory, announced that he and Dale Frail, from the National Radio Astronomy Observatory in New Mexico, had found two planets around the pulsar PSR 1257+12, which is some 1,600 light years away. This pulsar spins very rapidly, 161 times per second. He had been observing this pulsar for some years, and found that the radio pulses varied in their arrival time by as much as 1.5 thousandths of a second! This pulsar travels along its orbit quite slowly, at only 0.7 metres per second (about 2.5 kilometres per hour, or a slow walking pace).

So, after Wolszczan's findings, we're pretty sure that there are planets around at least one pulsar, but there's no way that life (at least, as we know it) could survive there.

Velocity

The sixth technique has found planets around stars like our own Sun. We know that a star and its invisible companion will orbit about their common centre of gravity.

In this technique, astronomers look for the change in velocity of the star as it sometimes moves towards us, and at other times moves away from us. They exploit the well-known Doppler effect, as used in police radar. The police use the change in frequency of the reflected radar beam to work out the speed of the vehicle — the higher the change in frequency, the higher

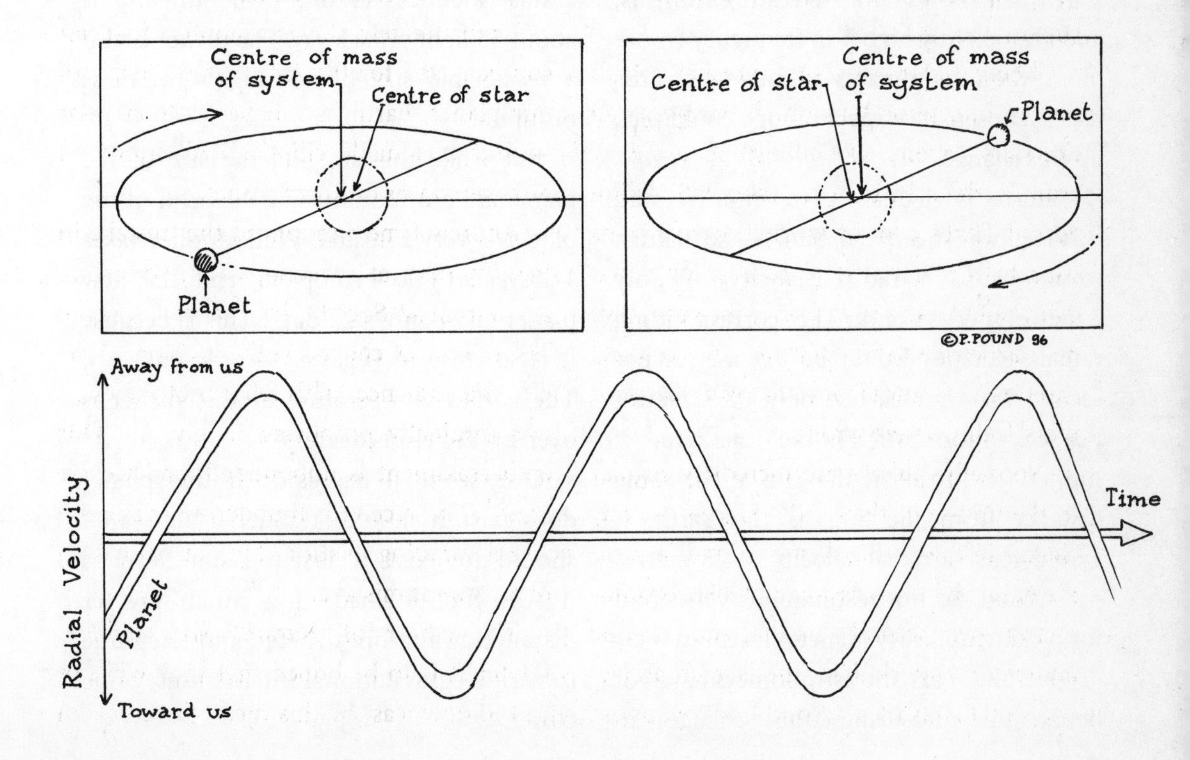

the speed. You can actually hear the Doppler effect in a siren. The frequency sounds higher as the siren approaches, and lower as it retreats.

Imagine that our Sun had Jupiter as its only planet. Jupiter would not revolve around the centre of the Sun. Instead, both Jupiter and the Sun would revolve around their common centre of gravity. Imagine that an outside observer was looking at our Sun, and was located in the plane of the Sun's equator. Sometimes, thanks to the mass of Jupiter, the Sun would appear to approach the observer at 12.5 metres per second, while at other times it would appear to retreat at 12.5 metres per second. The astronomers call this number the 'radial velocity'. This radial velocity of our Sun (thanks to Jupiter) would cause a change in the frequency of light by only three parts per million!

The radiated light from the star would get very slightly lower in frequency (more red) when the star was retreating from us, and very slightly higher in frequency (more blue) when the star was advancing towards us. A very sensitive spectroscope could pick up this change.

Saturn is lighter than Jupiter, and further away from the Sun — so Saturn would cause a radial velocity of only 3 metres per second. The corresponding radial velocities would be 0.3 metres per second for Uranus, but only 0.09 metres per second for Earth. On the other hand, if Jupiter were to orbit the Sun at the same distance from the Sun as the Earth, it would cause a radial velocity of 28 metres per second. So this technique is very good for picking up heavy planets close to stars.

Over the years, the astronomers have got better at measuring the change in frequency, which lets them work out the radial velocity of the star. In 1980, astronomers could measure 1,000 metres per second of radial velocity, and by 1992 had improved this to 20–25 metres per second of radial velocity. But by mid–1995, Geoffrey Marcy, from San Francisco State University, and Paul Butler, from the University of California at Berkeley, were able to consistently measure down to 3 metres per second.

But even though they had the most sensitive equipment on Earth, they were beaten to the punch.

On 6 October 1995, Michel Major and Didier Queloz, from Geneva Observatory, reported that they had used this technique to 'find' a planet orbiting around a star called 51 Pegasus. They had begun their search in September 1994. They had spent over a year, scanning through more than 142 stars every two months, before they had hit the jackpot: 51 Pegasus, which is almost identical to our own Sun and is about 55 light years away. The planet is also a 'reasonable' size, being roughly the mass of Jupiter.

But while the star and the size of the planet are not too unusual, the orbit of the planet is very unusual. While Jupiter takes 11.9 years to do a complete orbit, this new planet takes only 4.22 days! This is because it is so close to Pegasus — only 0.05 AU. The surface temperature of this planet is over 1,000 Celsius degrees!

One scientist has said: 'If this planet does exist, it shouldn't!' Such a big planet should not exist so close to a star. Perhaps it was first formed at a much greater distance away from the star, and later, due to various gravitational interactions, came closer.

NOT NUMBER 1, BUT 2 AND 3 . . .

Geoffrey Marcy and Paul Butler were almost the first to find planets orbiting around a star like our Earth's star. Their equipment was able to detect down to 3 metres per second velocity change in a distant star (as the star moved towards and then away from us). It was certainly more sensitive than the equipment of Major and Queloz (13 metres per second), but Major and Queloz were first. Major and Queloz got the honour of discovering the first planet, outside our Solar System, orbiting a star like our own Sun.

Marcy and Butler had been surveying 120 likely stars since 1987, and they had the data 'in the can'. But they had decided not to examine the raw data early in the project, because they felt that any planets big enough to detect easily would take 10 or so years to complete an orbit. They certainly did not think that a planet as large as Jupiter could be closer to the star than Mercury is to our Sun. (Neither did anybody else, for that matter.)

As soon as Michel Major and Didier Queloz announced 'their' planet around 51 Pegasus, Geoffrey Marcy and Paul Butler used their equipment to confirm the finding — and they found that it was real, at 51 metres per second! It was easily within the range of detection of their own equipment.

Then they began analysing the raw data to find their 'own' stars. Fellow astronomers lent them six computers to analyse the data. It took them two months, running the computers day and night, to get through 60 of the stars in their records. But on 30 December 1995, within three minutes of starting on 70 Virginis B, they knew that it had a planet. It had a velocity change of 311 metres per second — ridiculously easy to detect. The data had been sitting on their hard disc for seven years!

47 Ursae Majoris B took a little bit longer than three minutes to tease out of their raw data. It had a velocity change of 45 metres per second! As Butler said: 'We've built an enormous sledge-hammer to crack a walnut.'

That was probably their best chance of ever getting into the astronomical *Guinness Book of Records* — and they blew it. Sometimes, that's just how the cookie crumbles!

THE PLANETS OF PULSAR PSR 1257+12

According to Alexander Wolszczan's analysis of pulsar PSR 1257+12, this pulsar had two planets. One, at a distance of 0.36 AU, took about 67 days to do a complete orbit, and was more than 3.4 times the mass of the Earth. The other, at a distance of 0.47 AU, took about 98 days for an orbit, and was more than 2.8 times the mass of the Earth. According to his calculations, here were two planets that were a bit heavier than our own Earth, but orbiting an impossibly strange star some 15 kilometres in diameter, but weighing some 1.4 times the mass of the Sun. The radiation blast from the radio beam would have sterilised any known Earth life form.

Luckily for Wolszczan, the orbit times of 66.6 and 98.2 days were not some easy fraction of an Earth year, or a lunar month. This made the orbit times much more believable to the sceptics. In fact, because they were very close to a 3:2 ratio, quite regularly (after three orbits of the 66.6-day planet, and two orbits of the 98.2-day planet), the planets would be very close to their original starting positions of some 190-or-so days earlier. This would lead to recognisable and predictable changes in the timing of the radio signals.

Wolszczan spent the next few years observing and, sure enough, the predicted changes happened! Since then, he says that he has found another two planets orbiting around the pulsar PSR 1257+12.

Even so, it was the first 'sighting' of a planet-sized object around a star very similar to our own Sun. The year 1995 will be remembered as the year of the discovery of the first planets orbiting around stars just like our own Sun.

In January 1996, Geoffrey Marcy and Paul Butler found two more planets spinning around stars that were fairly similar to our own Sun. Both of these stars are quite close (35 light years away), they are each visible to the naked eye, and they each have a large planet orbiting at more 'reasonable' distances from their parent star.

One star, 70 Virginis B, is just a few hundred degrees cooler than our own Sun. Its planet is about eight to nine times heavier than Jupiter. It has a fairly eccentric orbit, which is a little unusual for a planet (if current theories of planet formation are anything to go by). Because it orbits at a distance of 0.43 AU, it takes just 116 days for a complete orbit. And the astronomers are very excited (as far as life-as-we-know-it is concerned), because the surface temperature — about 83 Celsius degrees — is compatible with water!

The other star, 47 Ursae Majoris B, is about three to four times heavier than Jupiter. Its orbit is almost circular. It orbits at a distance of 2.1 AU, and so its year is about 1,100 days (3.2 years) long. The surface temperature is estimated to be about *minus* 80 Celsius degrees.

Then, in April 1996, Geoffrey Marcy announced the discovery of another Jupiter-sized planet orbiting the star HR 3522 (also called Rho Cancri), some 40 light years away. A few weeks later, he announced the discovery of another planet orbiting the star Tau Bootis. In June, George Gatewood announced a Jupiter-sized planet orbiting the star Lalande 21185, which is the fourth-closest star to the Sun.

Discoveries

So now, in 1996, we have a few planets that definitely exist around other stars. Let me sum up the situation.

Two planets, a bit bigger than our Earth, orbit a bizarre star so shrunken, that a teaspoon of this star's substance weighs more than Mt Everest! This pulsar is about 1,600 light years away.

We have also 'found' six very large planets, each bigger than Jupiter, each orbiting a star very similar to our own Sun. These six parent stars are within 60 light years of us. Plus, there is also good evidence for a planet disturbing the disc around Beta Pictoris.

And finally, we have found that in the Orion Nebula, some 1,500 light years away, roughly half the stars we have closely observed have dust discs around them. We think that these dust discs will probably turn into solar systems in the next 20 million to 100 million years or so. We have also found about half-a-dozen dust discs that appear to have 'gaps' in them. We think that these gaps might be caused by a planet-sized body orbiting in the dust disc, sweeping up the gas and dust, as it grows bigger.

In our galaxy alone, there are some one billion stars that closely resemble our Sun. Going by the odds, there are probably 10 million stars, similar to our Sun, that have planets.

But what we have not found is an Earth-size planet, orbiting a star similar to our own Sun. And that is why, all around the world, astronomers are planning new and better telescopes, looking for this 'Holy Grail'. In the words of Paul Butler: 'I predict that within one year, more extrasolar planets will be known than there are planets in our Solar System. And that only counts normal stars. Pulsar planets are a whole other class.'

REFERENCES

Bruning, David, 'Lost & found: pulsar planets', *Astronomy*, June 1992, pp. 36–38.

Caillault, Jean-Pierre, 'The new stars of M42', *Astronomy*, November 1994, pp. 40–45.

Davies, John, 'Searching for alien earth', *New Scientist*, no. 1977, 13 May 1995, pp. 24–28.

Fienberg, Richard Tresch, 'Pulsars, planets, & pathos', *Sky & Telescope*, May 1992, pp. 493–495.

Mayor, Michel & Queloz, Didier, 'A Jupiter-mass companion to a solar-type star', *Nature*, vol. 378, 23 November 1995, pp. 355–359.

Morrow, Lance, 'Is there life in outer space?', *Time*, 19 February 1996, pp. 58–65.

Walker, Gabrielle, 'Seven planets for seven stars', *New Scientist*, 15 June 1996, pp. 26–30.

Wolszczan, Alexander, 'Confirmation of Earth-mass planets orbiting the millisecond pulsar PSR B1257+12', *Science*, vol. 264, 22 April 1994, pp. 538–542.